Prepper's Water Survival Secrets

Cost-Effective Water Filtration, Purification & Storage Strategies for Your Family to Thrive in Any Crisis

Hadassah Lewis

Contents

Introduction

Near Lake Chad in central Africa stands the town of Bol. Once a flourishing village on the shores of the lake, Bol now stands nearly a hundred miles north of Lake Chad's life-giving water. Even though the town hasn't moved, the lake has receded. Since the 1960s, Lake Chad has shrunk by over 90%, a devastating turn of events for the locals who have lived along its shores for generations. (*Finding Hope in the Face of Drought in Bol, Chad*, 2023) This harsh reality has its roots in climate change and desertification.

You might expect the reed huts of Bol to be deserted, for its people to be packing up and moving on to look for new lands to grow their crops and graze their cattle. Instead, the residents of Bol have worked to develop innovative water solutions, digging cisterns and implementing drip irrigation systems that enable them to grow crops in even the driest conditions. The community has come together, pooling resources and knowledge, to give their village a chance to thrive. Land reclamation projects are underway, taking back the land the desert has stolen.

Worldwide Water Inadequacy

This problem is not just happening in Bol; around the world, water scarcity is a real problem. It is estimated that in Africa alone, over 500 million people live on land that is experiencing desertification. (Schleeter, 2022) And even when there are adequate amounts of water, it is often polluted and unsafe for drinking. UNESCO estimates that one in nine people worldwide only has access to drinking water from unsafe sources. Also, 90% of sewage in developing countries is dumped untreated into water sources. Factories worldwide dump an estimated 200-300 megatonnes of waste into bodies of water each year (*The Global Water Quality Challenge & SDGs*, 2016). Considering all these facts, it's easy to see why our global water picture can seem grim.

For those of us living in developed countries, this global reality can be easy to ignore. We turn on our taps, and clean water flows from them. But the fact that you've picked up this book shows that you're ready to look at the reality of the situation - climate change, natural disasters, and various other catastrophes can happen anytime, anywhere, and to anyone. These events could affect our water supply without warning. But your picking up this book also shows that, like the people of Bol, you are ready to join a growing group of people who are prepared to pool knowledge, our most incredible resource, and gain the tools to thrive no matter what challenges we face.

Water Emergency Preparedness

You may have questions about how this could be possible, however. Anxiety over what kinds of natural or social upheavals we may face in the future can leave you feeling like you have more questions than answers.

- Where can I find practical solutions that will really work?
- How can I separate trustworthy information resources from clickbait and sensationalist propaganda?
- How much will it cost to prepare my family for a potential water shortage?
- Where can I learn the technical details of water purification, storage, and sourcing without spending hundreds of hours poring over books and websites for the most helpful and relevant information?
- Is it truly possible to be prepared and equipped with the knowledge and skills to address water scarcity challenges in a disaster or emergency?
- Do the skills needed in a water emergency have any real-life application in the world I live in today?

Please be assured that these questions have answers that can put your mind at ease. In this book, the hard work has been done for you. I have included information from reliable sources and practical, proven solutions. Reading this book and applying what you learn will equip your family for sustainable living today while preparing for a potential ongoing emergency water shortage in the future. I've combined the old, tried-and-true methods with a newer, modern approach to emergency preparedness, creating a well-rounded and practical guide for surviving water scarcity.

Imagine the peace of mind you'll feel when you have implemented my 31-Day Emergency Plan for water shortage preparedness. You'll know your family is ready to not just survive an emergency, but also to thrive in it. Not only that... you'll also understand water purification techniques and how to source water in an emergency, and you'll be able to overcome any challenges you may face along the way.

In addition to extensive research, I've also drawn from my personal experiences living in the Amazon jungle and growing up in a rural North American farming community. In these contexts, I've seen the life-giving beauty and the perilous hazard of water. On a typical afternoon in the jungle, while friends and I escaped the scorching heat by swimming in the river, my sensation of being in a tropical paradise was suddenly taken away by a shriek! A seven-year-old girl felt the river's strong undercurrent and panicked to the point she was no longer able to swim. At risk of drowning or being swept away, she called out for help. I was able to swim out to her and tread water to keep her afloat while her older sister and cousin swam to a nearby boat and rowed out to us. Thanks to teamwork and quick thinking, we brought the little girl safely back to shore, terrified but unharmed. However, we were all a bit shaken by the experience and by the awe-inspiring power of the river's water. As humans, we live in a delicate balance with this precious resource, where too little or too much, or even an unfortunate encounter, can mean death.

Work Out a Strategy

Just as my teenage friends and I helped that little girl, pooling our energy, resources, quick communication, and planning skills, my desire is that this book would help you. In the following chapters, I have compiled information on collecting, conserving, filtering, purifying, and storing water. Instead of being swept away in the overwhelming tide of information (and misinformation) surrounding water preparedness, I want you to be able to confidently implement a strategy that gives you peace of mind no matter what happens. If you apply the knowledge here, it may even save your life and equip you to save the lives of people you love.

But even if you never face that level of water emergency, you will benefit from learning these strategies and techniques. Wouldn't it bring tremendous freedom for you and your loved ones to develop a plan and know that you are currently using your water resources wisely and well? Come with me as we journey down the path of sustainability: we will learn about water conservation, potential sources of clean drinking water, different ways to purify water, and how hydration leads to good health and a healthy, thriving life.

1

Water's Worth: The Lifeline in Crisis

"Water is a commodity not by any means to be found everywhere...When found, it is more than likely to be bad, being either from a bitter alkaline pool, or from a hole in a creek, so muddy that it can only be called liquid by courtesy." — *Theodore Roosevelt*

Understanding Water's Role in Survival

If you've read my other books, you know I was raised in a home where natural remedies were our normal go-to. I was raised understanding the importance water had for health. If I went to "Doctor Dad" with a headache, I knew his first response would be to ask how much water I had drunk that day. He understood that my brain could not operate at its best without adequate water. Even slight dehydration can cause headaches, brain fog, confusion, and inattentiveness. This is because water accounts for three-fourths of brain weight, meaning your brain is 75% water. (Migala, 2015)

Water has many lesser-known health benefits and the well-established benefit of avoiding dehydration. Most of us know that adequate water helps lubricate our mouth, eyes, and nose, but did you know that it also lubricates your tissues, joints, and spinal cord? (Velayutham, 2017) In addition to that, it helps regulate body temperature, flush waste products, and carry oxygen to cells. Not only that, but specific vitamins such as Vitamin C, Vitamin B-12, Riboflavin, Folate, and several others cannot be broken down and absorbed without adequate water intake. (Wergin, 2022) Water is also vital to your body's immune system. Plasma cells, the cells in your blood that naturally produce antibodies, are 90% water, meaning that your immune response will suffer without enough water to drink. (Migala, 2015)

With all of this research showing the importance of water to human life, it seems as though making sure it's readily available to everyone should be a worldwide priority. Unfortunately, this isn't always the case.

Water Challenges of Today

Survival wisdom tells us that the human body can, on average, survive 3-4 minutes without oxygen, 3-4 days without water, and 3-4 weeks without food. (Peterson, 2023) Water also has many other uses that humans rely on besides drinking, including agriculture and hydroelectric energy production. A drought can affect all of these areas, creating problems that have the potential to seriously disrupt society.

The Brazilian Water Shortage

The years 2012-2015 represent what many have called the worst drought in the history of Brazil. During these years, seventeen of the biggest reservoirs in Brazil hit the lowest levels of the century. Over four million residents of southeastern Brazil were affected by power rationing and rolling blackouts as hydroelectric power plants in the region faltered, causing blackouts. Agriculture suffered devastating losses as sugar cane and coffee beans dried up in the fields. A single cane processing plant reported 250 lay-offs due to cane shortages caused by the drought. (Watts, 2017)

In poor areas of São Paulo, the nation's largest city, residents' taps were turned off for months at a time. Lack of clean water, rising unemployment, and unreliable access to electricity during a sweltering summer were the perfect recipe for social unrest, culminating in protests and rioting throughout the city.

What caused such a multi-year crisis, and more importantly, can we stop it from happening again? While scientists are divided on how significant a part each contributing factor played in the water shortage, they all agree on several factors that combined to create this devastating event.

- **Climate Change**—Due to increased warming, the cycles of El Nino and La Nina, which bring predictable rainfall to this area, are becoming erratic. Warmer temperatures also increase the need for water. This increased need affects not only drinking water but also extra irrigation and increased electrical consumption for air conditioning.

- **Overpopulation**—Areas of Brazil that had experienced rapid population growth were most affected. During the last 70 years, the population of São Paulo has grown from less than 3 million to over 22 million people (*Sao Paulo, Brazil Metro Area Population 1950-2022*, n.d.). This population explosion created a need for significantly greater resources.

- **Insufficient Infrastructure**—Long-term planning and maintenance of crucial infrastructure had been a struggle for the Brazilian government. As politicians focused on more popular and visible projects to try to win elections, upgrades on critical systems and urban plans were put aside.

- **Amazon Deforestation**—Tropical forests, such as the rain forests of the Amazon River basin, add moisture to the air. As these forests continue shrinking, the air dries out, resulting in less rainfall in the surrounding areas. This ongoing issue was a significant contributor to the drought in Brazil.

As you can see, several of these contributing factors were avoidable. Not all water emergencies can be prevented entirely. Still, their frequency, duration, and consequences can be lessened in many cases if people learn to mitigate the man-made problems that contribute to them. (Slater, 2019)

Water Stress

Water stress occurs when the supply of safe, usable water is insufficient to meet an area's demands. Approximately half the world's population lives in areas lacking adequate drinking water, sanitation services, or both. (Klobucista & Robinson, 2023) This can be due to physical water scarcity or inadequate infrastructure for the population. Water stress can affect diverse areas, ranging from places such as the Middle East, where physical water scarcity is a problem, to Ukraine, where war-damaged infrastructure is unequipped to provide water for locals. We usually think of this as a problem that affects economically unstable countries, but first-world countries can struggle with infrastructure shortages as well. In 2015, the city of Flint, Michigan, experienced water stress when the local water supply was found to be contaminated with lead due to old lead pipes that were still being used to deliver water to homes and businesses. (McBride & Berman, 2023)

Four out of every ten people on earth are currently experiencing water shortages, and experts predict that for every 1° Celcius that the earth's atmosphere warms, renewable water resources will drop by 20%. (Klobucista & Robinson, 2023) This will impact both drinking water and agriculture since estimates show that up to 70% of freshwater resources are used in food production.

Safety Concerns in Drinking Water

Clean, safe drinking water is vital to human life, yet billions live without access to it. According to UNICEF, about 350,000 children under the age of five die each year from diseases caused by a lack of safe drinking water. (Gonzalez-Cebrian, 2019) One factor contributing to this tragedy is that 80% of wastewater worldwide goes back into local ecosystems without undergoing any treatment. Untreated water can carry diseases like cholera, polio, hepatitis A, typhoid, and diarrhea. In addition to pathogens, untreated water may contain parasites. Gathering this unsafe drinking water often involves hours of waiting in line at community faucets or trekking to the nearest body of water, usually a river or lake, to collect water that may contain animal feces or chemical runoff.

Even in places where drinking water has been treated and should be safe, routine tap water testing often reveals unwanted ingredients in the water supply, usually in the form of chemicals. Chemicals such as lead, arsenic, mercury, fluoride, and chlorine have been found in drinking water in developed countries worldwide. Pesticides can also make their way into groundwater and are often not removed at treatment plants. Not only that, studies have shown that 95% of tap water in the USA contains microplastics. Microplastics are any plastic particles under five millimeters in diameter. Human consumption has been shown to disrupt our metabolic and cellular function as well as increase inflammatory response and put us at an increased risk of cancer. I'm guessing you would never eat a credit card. But would you have imagined that the average human consumes enough microplastics each week, mainly through drinking water, to create a whole credit card? This fact was revealed in a recent study. (Liu, 2022)

Common Misconceptions About Water Challenges

"In western Zambia, small villages dot the landscape near the Zambezi River. As our vehicles bounced over the dirt tracks, we were loaded with gallons of clean water, approximately one gallon per person for each day we planned to be there. We would use it to drink, cook with, wash our dishes, and brush our teeth. We would gather water from the river to wash our bodies and clothes, just like the locals." - disaster relief worker.

In situations like this, I've heard people say the only reason teams need to carry their own water is because they aren't used to the pathogens in the local water. But as we went from village to village, the truth emerged. Villages that had accepted the help of organizations with access to heavy machinery capable of boring holes deep into the earth to dig wells stood out from those whose leaders rejected outside help. In these villages with wells, people were healthy. They took their jugs to the well every day, pumping them full of clean water from the earth. In the villages without wells, more people were sick. They told stories of young children who had died of diarrheal diseases and of mothers killed by crocodiles while gathering water from the river. There were people with disfiguring growths, which we weren't seeing in the villages with the new wells. One of the leaders who had refused a well asked how long it would be before he could get one for his village. He had seen the difference that clean water made in the villages across the river where the wells were.

It's a misconception to think that human bodies get used to drinking the pathogens found in untreated water. That's just one of the myths I'd like to take a moment to debunk because knowing the truth is the first step to understanding our part in the solution.

Myth #1 - Any Water is Better Than No Water

Having access to unsafe water can be just as deadly as having access to no water. While our bodies can live between three and four days without water, water-borne illnesses such as cholera can kill in as little as twelve hours. (World Health Organization, 2022)

Myth #2 - People Shouldn't Live Where There is Not Enough Water to Support Them

Moving to areas with fewer people or more plentiful resources is a luxury. Suppose the resources in a specific area have diminished, overpopulation has occurred, or infrastructure has aged to the point that the water supply is no longer safe. Relocation takes resources that people in this situation may not have. (Adesina & Beadle, 2020)

Myth #3 - Rainwater is Safe to Drink Because It Is Not Polluted Yet

Although capturing rainwater can be a great way to conserve water, it is not considered safe to drink. Rainwater can contain the same pollutants, chemicals, and fecal matter bacteria as standing water. Rainwater can be used freely for many daily activities: watering plants, laundry, flushing toilets, and bathing. This can be especially helpful when water resources are low. However, rainwater should only be consumed if it has been thoroughly purified. (CDC, 2019)

Myth #4 - Water Has an Expiration Date

Water doesn't expire. However, water that is contaminated with bacteria or algae before storing will have an increasing amount of contaminants as it gets older, leading to the belief that water can "go bad." (Emergency Essentials, 2013)

Myth #5 - Non-Potable Water Can Be Used to Wash Dishes

It is never a good idea to wash dishes or plants that will be eaten raw in water you wouldn't drink. The same bacteria you avoid by not drinking the water will get onto your food or dishes in the washing and rinsing process. Therefore, it's vital to always use drinking water for these activities. (Adesina & Beadle, 2020)

Acknowledging But Not Afraid

Learning about the issues facing our world's water supply can feel overwhelming. But I'm not here to report doom and gloom! Knowing the scope of the problem is an essential first step in making sure that you're prepared to meet it. Throughout this book, we will focus not on the issues with the water supply but on making sure that your family will be provided for in this vital area, no matter what.

Understanding Your Water Needs

So, how much water do you need to function at optimum brain and body performance levels? The answer isn't as straightforward as health gurus often make it sound. Depending on age, size, and gender, we all need different amounts of water. (Karanja, 2023)

Total Water Consumption Needs

Age Range	Male	Female
Less Than 1 Year	Up to 4 Cups Daily	Up to 4 Cups Daily
1-2 Years	5 Cups Daily	5 Cups Daily
2-3 Years	5.5 Cups Daily	5.5 Cups Daily
4-8 Years	7 Cups Daily	7 Cups Daily
9-13 Years	9 Cups Daily	8 Cups Daily
14-19 Years	10.5 Cups Daily	8.5 Cups Daily
20 and Up	11 Cups Daily	9 Cups Daily

Of course, these are averages. Not only that, but a pregnant woman should plan to increase her water intake by about a cup a day, and lactating women need two to five extra cups of water per day. (MacGill, 2021)

Total Water Consumption vs. Drinking Water Consumption

Our bodies can source approximately 20% of our water needs from our food intake. This can be higher if you eat many foods with a high water content, such as fresh fruits and vegetables. It will be lower if you drink a lot of caffeine or alcohol. Both caffeine and alcohol are diuretics, so consuming excessive amounts will increase the amount of water your body needs.

Looking at the chart above, you can determine how much water you should drink daily. Increase the number of cups if you have a strenuous exercise routine or live in a hot climate. Both variables cause you to lose more water via sweat each day. Also, make any necessary adjustments for pregnancy, lactation, or any dietary attributes that may affect your water intake needs. Once you know your total water consumption needs, you should drink roughly 80% of that amount.

More - Not Always Better

While it is dangerous to consume too little water, having too much can also be harmful. Excess water intake, over seven gallons daily, can cause your body to flush out too much salt. (Gorin, 2022) This can lead to muscle cramping, fatigue, brain swelling, lung congestion, and even death. (MacGill, 2021) Fortunately, this rarely happens, and when it does, it usually only affects people with diabetes, endurance athletes, or those taking medications that affect water retention in the body.

The Prepper Mindset

Now that we've explored our need for water and the scope of the global crisis, it's time to discuss how we can ensure our families are safe and assured of a reliable water supply no matter what happens around us. Preparing for a potential water crisis does not have to be all-consuming or focused on catastrophe. As we've seen, these events can happen anywhere and to anyone. Preparing for them is a good-sense decision.

The prepper mindset has three crucial components: preparedness, sustainability, and resourcefulness. Harnessing prepper's knowledge and skills can help you increase your resilience when facing a crisis.

Emergency Preparedness

Knowing that your family is prepared for an emergency is one of the main elements of the prepper mindset. By definition, emergencies are usually unexpected. Natural disasters such as earthquakes, flash floods, and tornados come with no warning. Other emergency situations, such as hurricanes, may include a small amount of preparation time, so you will be fighting for limited resources with everyone else trying to prepare. Preppers know it is cheaper and easier to find resources ahead of time. Once you have developed a plan, you can have peace of mind knowing that your family is safe.

In addition to peace of mind, being prepared ahead of time for an emergency can save lives. Maintaining a store of adequate food and drinkable water can be the difference between life and death for family members and pets. Knowing safe evacuation routes can prevent you from driving into potentially unsafe conditions. Preppers recognize that the time and resources it takes to be ready ahead of time will be well worth it if an emergency strikes.

Resourcefulness

Resourcefulness is the ability to deal skillfully with unfamiliar situations and devise ways to meet the challenges those situations pose. (Merriam-Webster, 2024) This is, in many ways, the essence of the prepper mindset. Resourceful people make choices by considering potential challenges and planning ahead for them. When I lived in the Amazon rainforest, I had to choose whether or not I wanted indoor plumbing. Initially, using the community showers and sinks seemed like the best idea. However, when I thought through the potential downsides of only having access to community showers, especially in the tropics, I decided to add a bathroom to the blueprint for my little home in the jungle. However, I decided against a kitchen sink. This turned out to be the right choice because washing dishes in the community sink under the mango trees while watching monkeys or chatting with friends was much more fun than doing that chore at home alone.

Resourcefulness when preparing for emergencies involves identifying your family's priorities and goals for preparedness needs. What types of natural disasters are most common where you live? How many people and animals would you be trying to keep safe in a disaster? Would it be wiser to leave your home or stay in an emergency? Preppers know that answering these questions and others like them will help them protect their family using the resources at hand.

Sustainability

It's no secret that the impact of humans on the environment is growing. Left unchecked, humanity will eventually need more resources than the earth can produce. Sustainability is the idea that each generation needs to provide for its needs without impairing the ability of future generations to do the same. (BluGlacier, 2020) Preppers understand that living sustainably will positively impact our environment, reduce pollution, and help society as a whole.

Living sustainably can help protect biodiversity and reduce deforestation and desertification. It can also enable you to live completely off-grid if this is one of your goals. Preppers know that using alternative energy sources, reducing waste, eating local, and limiting our water use can positively impact the world and improve our personal quality of life.

Quiz: How Sustainable is Your Lifestyle?

Choose the answer that most closely resembles your current lifestyle choices to get an idea of how sustainably you live.

Diet Choices

a) I mainly eat plant-based or vegetarian meals.

b) I occasionally eat meat and fish but mainly eat plant-based.

c) I eat meat and fish regularly.

d) I consume meat or fish at every meal.

Transportation Habits

a) I mostly walk, cycle, or use public transport.

b) I drive a fuel-efficient or hybrid car.

c) I drive a regular car occasionally.

d) I drive a regular car or a gas-guzzling vehicle every day.

Waste Reduction

a) I always use reusable bags, bottles, and containers and try to buy products with minimal packaging.

b) I often use reusable items but sometimes use single-use plastics.

c) I rarely think about packaging and often buy single-use plastics.

d) I mainly use single-use plastics and don't recycle.

Energy Consumption

a) My home uses renewable energy sources, and I'm conscious about reducing energy waste.

b) I turn off appliances when not in use but haven't switched to green energy.

c) I occasionally forget to turn off lights and appliances.

d) I often leave lights and appliances on, even when not in use.

Water Use

a) I take short showers, turn off the tap when brushing my teeth, and use water-efficient appliances.

b) I'm conscious about water use but could improve in some areas.

c) I take long showers and occasionally wastewater.

d) Water conservation is not a priority for me.

Clothing and Shopping Habits

a) I buy second-hand, sustainable brands or only buy what I truly need.

b) I occasionally buy from fast fashion but try to choose sustainable options.

c) I often buy from fast fashion but don't shop excessively.

d) I frequently shop and prioritize trendy items over sustainability.

Results Calculation:

Assign points for each answer:

a) = 4 points

b) = 3 points

c) = 2 points

d) = 1 point

Total Score:

- **20-24 points:** *Highly Sustainable* - You're making excellent choices to minimize your ecological footprint. Keep it up!

- **14-19 points:** *Moderately Sustainable* - You're on the right track. With a few changes, you can significantly reduce your environmental impact.

- **8-13 points:** *Somewhat Sustainable* - There's room for improvement. Consider making more eco-friendly choices in your daily routine.

- **6-7 points:** *Not Very Sustainable* - Your habits are having a considerable environmental impact. Consider researching ways to adopt a more sustainable lifestyle.

Gaining and Implementing Knowledge

Preparing for a water emergency in advance is a powerful way to protect the people you love. We've seen that an adequate clean water supply can make the difference between life and death, especially since climate change, deforestation, and growing populations continue to increase the likelihood of water scarcity. Examining pervasive myths about water and water shortages has increased our awareness of facts that can save lives. We've also learned a bit about having a prepper mindset: being prepared, resourceful, and sustainable. This can help us be part of the solution instead of the problem when an emergency strikes.

Chapter 2 will discuss where to find water and how to collect it. Learning to tap into the natural water sources in our local environment allows us to be prepared, whether or not we have space to store large amounts of water. So, let's continue our journey, where knowledge is the power to prepare for an emergency before it occurs.

2

Harvesting Nature's Bounty

"Nature is not a place to visit; it is home." — *Gary Snyder*

DID YOU KNOW THAT 70% of the earth is covered in water? (Water Science School, 2019) With that much water, it is hard to believe that water shortages are common. But they are. However, knowing how to find usable water is a skill that has allowed humans to thrive for centuries. In this chapter, we will tap into wisdom from the past so we can continue to thrive even in difficult situations.

Rainwater Collection

Collecting and using rainwater is an easy way to begin working toward water self-sufficiency. Water can be collected from any waterproof surface. A rainwater collection system can be extremely simple or quite complicated. Once your system is set up, you will have access to free water; this water is already usable for irrigation, flushing toilets, and laundry. It can also be purified for drinking and washing dishes.

History of Collecting Rainwater

For thousands of years, humans built cisterns to catch their most precious resource... water. The oldest cisterns ever found are from about 4000 B.C. in the Levant region of Southwest Asia.

Two thousand years later, the great civilizations of Asia began carving great barrels out of rock, with a system of stone gorges running through the city to release water in case of drought (Lindholm, 2019). The largest ancient rainwater collection cistern is in India. Dating from 300 B.C., it collected rainwater from the roof of the Brihadeeswarar Temple. This cistern is nearly ten miles long and holds over ten billion gallons. (Microfarm Garden, 2019)

Benefits of Collecting Rainwater

There are many motivations for joining the long list of civilizations that have collected rainwater throughout history. Although storing rainwater is no longer a matter of immediate survival, we can still benefit in many ways from this time-tested way to be prepared for water shortages.

- It provides a backup water supply for emergency use.
- Rainwater is entirely free and relatively clean.
- It contributes to self-sufficiency.
- It helps conserve water.
- Collecting rainwater stops excess runoff from causing erosion.
- It gives you an autonomous water supply.
- Rainwater collection is environmentally responsible.
- Non-chlorinated water is better for watering the landscape.
- Using gathered rainwater when possible can decrease your utility bills.

These are a few of the many benefits of rainwater collection for both you and the earth (Maxwell-Gaines, 2018). Collecting rainwater has been practiced throughout history because it has so many advantages and so few disadvantages.

Rainwater Collection in Practice

The most common rainwater collection systems have always been those that collect water from the roof of a building. Water is run through a simple filtration system, such as a net or screen, and stored in a closed container to avoid evaporation and contamination. The design of these systems can vary considerably. (Innovative Water Solutions, LLC, 2023) As technology advances, the ability to set up rainwater collection systems that filter water and produce drinkable tap water is growing. The three main types of rainwater collection systems are rain barrels, dry systems, and wet systems.

Rain Barrels

The simplest type of rainwater collection is a rain barrel, where you use your existing gutter system to collect the rainwater. Then, you install a collection barrel at the bottom of one or more of your gutters. Collection barrels can range from one bought from a hardware store to a repurposed barrel, drum, or trash can. A few points to consider when setting up a collection barrel are material, size, and placement.

- Material: Barrels for your rainwater collection system can be made of plastic, ceramic, stone, concrete, or wood.

- Wooden barrels are best if made from oak, cypress, or another rot-resistant wood. These barrels need more upkeep than other types; they should be treated for wood preservation every few years. They are considered aesthetically pleasing but need periodic sanding and restaining to keep that appearance up. They usually hold about 60 gallons, which makes them among the largest rainwater barrels for sale. This is also a sustainable choice since it limits plastic use at home. These barrels are sturdy and less likely to be knocked over in a storm, even if they are predominantly empty. However, wooden barrels need ventilation to avoid rot, so you need a mosquito control method to prevent mosquito larvae from growing in your rain barrel. (Westerlund, 2023)

  - Stone, concrete, and ceramic barrels can be used to collect rainwater. These are decorative and often considerably smaller than wooden or plastic barrels due to their weight. They are usually sealed on the inside or made of non-porous materials. These tend to be more expensive than wood or plastic, but they can be an excellent choice for a small garden. (Chapple, 2021)

  - Plastic is by far the most common material used for rainwater barrels. This is due to its inexpensive nature and long lifespan. One downside to these is that plastics may leach chemicals into your water supply unless you get food-grade plastic barrels. Looking at the recycling number can tell you if a barrel is food-grade. Recycling numbers 1, 2, 4, and 5 are food-grade plastic, meaning they won't leach chemicals into the water even if you store it long-term. This may or may not be a problem for you, based on how you plan to use the water in the barrel. For example, watering your lawn with water that has been kept in a non-food-safe barrel would be less of a problem than watering your vegetable garden. Plastic barrels are also light, making them easy to set up. They also usually come completely ready to install, making them an excellent choice for beginners. (Alderson, 2021)

- Size: Most households will need 40-80 gallons to make a difference in their water consumption. There is a formula to calculate how many gallons you could collect off your roof.

  - 1" of rainfall x 1 sq ft = 0.623 gallons

  - Find the average yearly inches of rainfall for your area.

  - Here's an example of the formula at work: If my home's footprint is 900 square feet, I install rainwater barrels at the end of every downspout, and I live in an area with an average of 50 inches of rain per year, I can collect 28,035 gallons of water from my roof on an average year.

- Placement: The easiest place to install rainwater barrels on a house with an established gutter system is at the end of your existing gutters near where you will need the water. For example, I installed a rainwater barrel at the downspout next to my garden. Rainwater barrels can also be installed where water runs off the roof of a shed or garage. Collecting rainwater is possible, even if you don't have a suitable rooftop from which to collect it. Any slanted surface will experience runoff during rain. This includes a slanted deck, a hill, or a driveway. Or you could build a tilted surface that feeds directly into your rain barrel. (Eileen, 2022)
    - One easy way to collect water into a freestanding barrel is to utilize a large plastic trash can with a domed lid. Turn the lid upside down and drill up to ten holes in it. Secure the lid to the trash can using zip ties. Near the bottom of the trash can, install a faucet, carefully sealing around the edges with silicone caulking to waterproof it. Set the trash can on cinder blocks in an open area, using bungee cords to secure it to the blocks. The upside-down lid acts as a funnel to trap rainwater and channel it down into the barrel through the holes in the lid. (Eileen, 2022)

Dry Collection System

A dry collection system is similar to a rainwater barrel. The main difference is that it's bigger. Usually using a tank larger than 100 gallons, dry collection systems are perfect in areas that experience a defined dry season and rainy season. This is because they hold much more water, enabling you to have more reserves for the dry season as long as you have been able to fill the tank during the rainy season. In addition to using gutters to direct water into a dry system, most people use pipes to direct water flow from areas of the roof that do not naturally flow in the direction of the water tank. This is called a "dry" system because the pipes dry out when there is no rain. The advantages of a dry system include a larger tank rather than a 50-60 gallon rain barrel, a simple setup, and less upfront cost and long-term maintenance than a wet collection system. (Firth, 2017)

Wet Collection System

In a wet collection system, rainwater from the roof is channeled through pipes underground and into a large water collection tank. This system is perfect for people who need to locate their rainwater tank at some distance from their house. In a wet collection system, underground pipes carry water from the gutters to the collection tank. Vertical pipes carry the water up to the tank inlet. As water enters the pipes, the water level rises in the vertical pipes until water spills into the tank. This is considered to be more secure than a dry system because the pipes are protected by being buried. As long as the underground pipes are waterproof, this system takes very little maintenance. It is, however, the most complex and expensive to install. This is called a "wet" system because the underground pipes hold water. One thing to pay attention to when planning for a wet collection system is that the tank inlet must be installed at a lower elevation than the lowest gutter that feeds into the system. (Innovative Water Solutions, LLC, 2023)

Creating a Rain Barrel Water Collection System

Before beginning a rainwater collection system, be sure to check state and local laws. Depending on your location, there may be permitting or other requirements for installing a rain catchment system. If you live in certain areas, you may receive government incentives. Be sure that whatever system you install meets the design requirements and that you are using the water for purposes approved in your area. (Wasson, 2022)

Making a rainwater collection system doesn't have to be complicated. Follow the step-by-step directions below to begin collecting rainwater in a DIY rain barrel system that uses your existing gutter system.

- Decide where you will put your rainwater barrel.
- Choose whether to buy a ready-made rain barrel or make one from a repurposed container. Suppose you plan to repurpose a barrel or trash can. In that case, you will need to buy a rainwater barrel spigot, flexible downspout adaptor, a one-inch garden hose adaptor, and silicone caulk, as well as have a drill with a one-inch saw bit and four-inch hole cutter for it.

- Begin by gathering your supplies. Decide which side of your barrel is going to be the front.
- Use your drill to cut a four-inch hole in the top of your barrel, about six inches from the back side of the barrel.
- Change bits, then cut a one-inch hole on one side of your barrel, about four inches from the top. This will be your overflow valve.
- Cut a one-inch hole near the bottom of the barrel on the front. Keep in mind that any water under this hole will not drain, so put it as close to the bottom as possible.
- Screw the garden hose adaptor into the overflow valve. Make sure that you thread it with the end that can connect to the garden hose on the outside.
- To install your rainwater barrel spigot correctly, follow the directions that came with it (RAINPAL sells great ones).
- Caulk around your overflow valve and spigot to avoid leaks.Install the four-inch round end of the downspout adaptor into the hole in the top of your barrel. Now you are ready to set your barrel up!
- Decide which downspout you will install the barrel under. Place a screen at the top of that downspout to filter the water entering your barrel.
- Place two cinderblocks side by side next to the downspout, leveling the ground if necessary. Stack two more cinderblocks on top of them, putting them at a 90° angle to the first ones to increase stability. Set the barrel directly in the center of the stacked blocks. (Neveln, 2022)
- Use a handsaw or tinsnips to cut the downspout to the required height. Attach the flexible downspout adaptor securely to the end of your downspout and wait for the rain to fill your barrel with clean rainwater. (*The Best RAIN BARREL for Less than $15, and Where to Find a Barrel*, 2007)

Groundwater Sources

Earth is called the Blue Planet because of the abundance of water it contains. Unfortunately, only about 0.3% of that water is usable by humans. (Mullen, 2019) Usable water is comprised of surface water and groundwater. Surface water is much easier to obtain, so it is used about four times as much as groundwater. (Mullen, 2019) Surface water is any readily available water source, such as lakes, rivers, or streams. Groundwater is underground water stored in aquifers. (Atlas Scientific, 2022) Groundwater is usually of higher quality than surface water since being stored underground protects it from most contaminants. It also represents the vast majority of the water that humans can use; 98% of the earth's available fresh water is stored underground. (Mullen, 2019) Since groundwater is more readily available than surface water, it makes sense to learn how to tap into this precious resource.

Finding and Gathering Groundwater

It's hard to collect groundwater if you can't find it. If you have property on which you live or intend to live, you may be above an aquifer. Below are a few clues that you can look for that suggest groundwater is nearby.

- A natural spring that does not dry up seasonally
- Sand or gravel in the bottom of a valley
- Places in a creek or stream where water flow increases with no visible explanation
- Vegetation that stays green even in dry seasons (Lone Star Drills, 2018)

The most precise way to know where and how deep to dig a well to access groundwater is to have a geological survey done. This will often need to be done for the permits required to drill a well. Check local and state regulations to find out whether or not your well needs a permit. There are three kinds of wells to choose from when attempting to access groundwater.

Dug Wells

This is the cheapest kind of well to make. As the name implies, a dug well is made by digging with a shovel into the soil that covers the aquifer. It is only an option when the groundwater is shallow enough to be reached by manual digging, and the soil is soft enough to dig through with no rock layers between the surface and the water table. (*Guide to Constructing a Water Well for Living Off-Grid*, 2022)

Driven Wells

Driven wells are made by driving a pipe into the aquifer to access the water. This pipe must have a screened "well point" on it to avoid sediment contaminating the water. These wells are generally between thirty and fifty-five feet deep. (Environmental Protection Agency, 2015)

Drilled Wells

Drilled wells are created by portable percussion or rotary drilling machines. These wells can be thousands of feet deep if necessary to reach deeper aquifers than either of the other types of wells. Because of the depth, drilled wells are the least likely type of well to be contaminated by surface activities. (Environmental Protection Agency, 2018)

Creating Your Own Driven Well

Supplies You'll Need

Be sure to get enough sections of pipe to reach deep enough to drive the pipe into the aquifer. Pipes for driving a well usually come in sixty-inch segments.

- Well point
- Pipe
- Couplings

- Drive cap
- Pipe joint compound
- Check valve
- Well pump
- Auger
- Sledgehammer, wooden mallet, or pile driver

How to Drive a Well

1. The first step in driving a well is knowing where to dig to hit an aquifer. You can do this using traditional methods such as water-witching, where a dowsing rod is used to find groundwater, or more modern techniques, such as conducting a geological survey of your property.

2. Once you know where to dig, begin by digging a pilot hole about two feet deep using the auger.

3. Screw the drive cap onto the well point and drive the well point into the ground using the sledgehammer until only about ten inches remain above ground. Remove the drive cap.

4. Apply a thin layer of joint compound to the threads on the end of the well point and the threading inside the coupling. Screw the coupling onto the end of your well point. Then, apply a thin layer of joint compound to the threading on the first piece of pipe that you will attach. Screw it into the other side of the coupling.

5. Screw the drive cap onto the end of the pipe before beginning to drive the pipe down into the ground.

6. Continue driving the pipe into the ground until about ten inches remain above ground. Repeat the coupling process.

7. Continue adding pieces of piping and joining them using the couplings and the joint compound until you hit the water. You will know that your well point is in the aquifer when it makes a hollow sound when you're driving it down with the sledgehammer.

8. For your pump to draw water efficiently, the well point should be about five feet into the water. You can measure how far into the aquifer the well point is by dropping a string with a small weight down the pipe until the weight hits the bottom of the well point. Draw the string up and measure how much of it is wet. Once the bottom five feet of the string is wet, your well is deep enough.

9. Screw the hand pump onto the top of your pipe, then pour a few cups of water into the top of the pump to help prime it. Work the pump until water comes out with every downward stroke. Remove the pump and install a check valve, remembering to seal the threading with joint compound. The check valve only allows water to flow in one direction, so it stops the water from flowing back down to the level of the water table. This makes pumping water easier and faster.

10. Reattach the pump, sealing it with joint compound to create a vacuum and ensure that it doesn't leak. Pump water from the well until the flow is clear. (MacDougall, 2011)

Testing Your Groundwater

Although groundwater is more protected than surface water, it can become polluted by landfill or septic seepage and pesticide or fertilizer runoff. This is true even though groundwater is filtered naturally, as gravity pulls it through layers of sediment and porous rock. This means that you should test groundwater intended for drinking, whether it is from a well or a natural spring. (Atlas Scientific, 2022)

You can buy testing kits for aquifers that test for pH, turbidity, saline content, nitrates, dissolved solids, and oxygen content (Atlas Scientific, 2022). These tests will help you ascertain whether the groundwater you have access to is safe to drink without further filtration or treatment.

Dew and Condensation

Condensation is the process by which water vapor in the atmosphere becomes liquid. This is what happens when water forms on the outside of a glass of cold water. Condensation happens naturally at night when air temperatures drop. As the temperature drops overnight, water vapor from the air condenses on outdoor objects, including leaves, porch furniture, or your car's windshield. The temperature at which this happens is called the "dew point." The dew point changes depending on the air temperature and the level of water vapor in the air, which is called humidity. Air with a higher humidity level has more water vapor, meaning more water will condense out of it. (Lancaster, 2024)

Gathering Dew

Water that condenses naturally outdoors with no external energy required is known as dew. With a bit of knowledge and effort, this water can be gathered and used. Although dew is considered to be drinkable water as long as you are confident that the dew has not been exposed to impurities on the substance you gather it from, it is recommended that, if possible, collected dew should be filtered through a paper towel or coffee filter as well as boiled for one minute to kill germs. (FEMA, 2017)

- Grass collecting involves harvesting dew off clean grass. Choose a patch of grass that animals don't have access to, free of impurities. When the dew has formed early in the morning, use clean, dry cloths to wipe it off the grass. When a cloth becomes saturated, wring the water into a bowl or pitcher. Repeat, changing the cloth when necessary, until you have enough water or have harvested from all of the grass. (Cunningham, 2023)

- Leaf collecting uses the same method as grass collecting, but from the leaves of plants in your yard or the wild. Ensure that the plants you collect from are not poisonous and haven't been sprayed with chemical pesticides. (Cunningham, 2023)

- Tarp collection involves harvesting dew from a tarp. To use this method, hang a tarp parallel to the ground. Add weight to the center of the tarp by attaching a washer to the underside with water-resistant adhesive, then cut a hole through the tarp that lines up with the hole in the center of the washer. Put a bucket under the hole. Water will drip into the bucket, and you can use clean cloths to wipe off any dew that stays on the tarp.

Gathering Fog

Another way to take advantage of atmospheric water vapor is when there is fog. Hanging a sheet or other thin fabric in the fog allows the vapor in the air to condense on the fabric. When the fabric is saturated, you can wring it out into a bucket or bowl (Cunningham, 2023). Once you have filtered the water and boiled it for one minute, this is considered safe drinking water.

Gathering Air Conditioner Condensation

One often-overlooked source of distilled water is your home's HVAC unit. Depending on the humidity where you live, your air conditioner generates between one and eighteen gallons of water per day. (Lancaster, 2024)

- Find the outlet where the condensate is currently draining. Usually, this is a PVC pipe sticking out of the side of your house with water dripping from it anytime your HVAC is running.

- Use connectors and lengths of PVC to direct the water away from the house to a garden or other area that needs irrigation. (Pickett, 2022)

- Alternatively, you can create a reservoir by digging out a space near your home that will hold a five-gallon bucket. Make sure the top of the bucket is covered by a screen to keep mosquitos out. Use PVC pipes and connectors to direct the water away from the house to your reservoir. If you live in a humid area where your HVAC will produce more water, you may want to make several of these bucket reservoirs. Using a PVC splitter will allow water to flow into all of them.

Water can be pumped out of these reservoirs using a hand pump. (Staff, 2022)

Plant-Based Water Extraction

One way to get water in an emergency is to extract it from the leaves of plants. The best plants for this process are large-leafed green plants in full sun, but other plants will also work. Just make sure you never extract water from poisonous plants!

You'll need a plastic bag and a string to extract water. Find a branch in the sun with plenty of healthy leaves. Tie the bag tightly at the end so no water vapor can escape. It will take several hours for water to condense on the inside of the bag in a usable quantity. The water should be strained through a small-holed filter or paper towel. After that, it can be considered clean and safe for drinking. (Malegowski, 2017)

Water Sources in the Desert

A desert is an area that gets less than ten inches of rain per year. If you live in or travel to a desert, it is critical to know what to do if you run out of water. One of the best options for securing water in a desert area is plant-based water extraction, which we talked about in the previous section. There are also other ways to find or make water in dry environments.

Even in the desert, dew falls overnight. Leave all receptacles out overnight to gather dew, then combine the contents into one cup early in the morning before the dew starts evaporating.

If there are cliffs, look at their bottoms for hollows where water may have collected during overnight rainfall. Water can be scooped from these hollows or soaked up with a cloth and wrung into a cup. Dry streams or riverbeds may also have hollows that still contain water, especially in rough, craggy terrain.

One way to increase your chances of finding water is by looking for vegetation or signs of animal and insect life. All plants and animals require water to live, so their presence means there is water, even if it is underground. Follow any vegetation to the lowest ground since water flows downhill. Once you are at a low point near vegetation, there are two ways of accessing water. (Sommerlad, 2018)

Create a Solar Still

Dig a hole in direct sunlight to about eighteen inches or the depth at which the soil begins to feel slightly damp. Fill the hole with green leaves from non-poisonous plants, leaving space in the center for a drinking cup. Cover the hole with clear plastic wrap. Pour sand or dirt onto the edges of the plastic wrap all the way around to seal the hole. Place a small stone on the plastic wrap directly on top of the cup. The plastic wrap should dip down in the center without touching the sides of the cup.

Water will condense from the leaves and soil and gather on the plastic wrap, dripping to the lowest point and falling into your cup. This will only work in full sun, so make sure that your hole will not be shaded as the sun moves. You can dig as many of these as you have the supplies for, and once you have collected the water, you can replace the leaves in the hole and repeat the process. (Jones, n.d.)

Dig a Collection Pit

Once the evening begins to cool off, dig a hole about twelve inches deep. If the ground feels damp at that depth, dig the hole to about twelve inches across. If not, go deeper, until the soil begins to feel moist, before widening the hole. Water will accumulate in the hole overnight, and you can soak it up with a rag and wring it out into a cup in the early morning before it evaporates. (Sommerlad, 2018)

In Summary

Water can be collected from our environments in various ways. Knowing what your options are for collecting water allows you to be prepared in the face of any potential water shortage.

Atmospheric Collection	Groundwater Collection	Overflow Collection	Emergency Collection
Rainwater	Dug Wells	Air Conditioner Condensation	Plant Water Extraction
Dew	Driven Wells		Solar Still
Fog	Drilled Wells		Collection Pit

Planning Ahead

When I lived in the Amazon, there was a broken bridge across the river near our village. The local authorities had built it a decade earlier, but their plan didn't account for the river's power. A working bridge would have made life so much easier! Instead of leaving our vehicles on the far bank of the river and ferrying supplies and people across in boats, we could have just driven ourselves across. That broken bridge became a reminder that any enthusiasm, ideas, and any old plan aren't enough. For long-term success, a plan must consider the reality of a situation.

This chapter has provided details on how to look around you and determine the most realistic ways to procure water in an emergency, taking your situation and needs into account. In the next chapter, we will begin to use this information to craft an emergency preparedness strategy tailored to your reality.

3

Crafting a Water Strategy

"We never know the worth of water until the well is dry." — *Thomas Fuller*

ANY STRATEGY FOR PREPARING for a water emergency must include several key components:

1. Figuring out where the water will come from.
2. Knowing how much water you need.
3. Deciding where to store excess water until it is needed.
4. Figuring out how to differentiate drinking water from non-drinking water.
5. Your plan must be personalized for seasonal changes in your area and your family's needs and habits.

Water Storage Solutions

Since Chapter 2 covered sources of water, we will begin with the next step in our water strategy - storage.

How Much Water to Store

Generally speaking, each person in your home needs about one gallon of drinking water each day. In addition, you need to store water for pets and livestock. A good rule of thumb for animals is to plan for one gallon of water per day for every hundred pounds of animals you have. (MacWelch, 2015)

This means an average-sized horse will drink ten gallons daily, and a cow will drink fourteen gallons daily. You will need a gallon of water per twenty chickens you have. Cats will need about a cup and a half of water per day, and dogs will vary based on size. Animal drinking water does not need to be the same quality as human drinking water, but should be free of physical and chemical contaminants. If you have livestock that drink from ponds, remember that salt and toxin levels may rise to dangerous levels during a drought as the ponds dry up. (Meehan et al., 2015)

In addition to drinking water, you need to store drinkable water for washing dishes, brushing your teeth, and rinsing produce that will be eaten raw. Non-potable water--that is, water that is not safe for drinking--can be your go-to for all of the following uses.

- watering plants
- washing clothes
- flushing toilets
- cleaning non-food-related items
- bathing
- handwashing

Having adequate reserves of non-potable water will free up your drinking water for its intended purposes. Irrigation is an important use of water during an emergency. In the summertime, a vegetable garden needs about half a gallon of water per square foot each day. In spring or fall, it requires half that much. (dePeyster, 2014)

Let's look at a couple of examples for figuring out how much water to store.

Rob and his family want to be prepared in case of a hurricane. Local authorities have suggested storing at least three days of drinking water before hurricane season. Rob plans to store five days of drinking water for his family. Rob has a wife and two children. They also have two small dogs and a cat. Rob needs one gallon per day for four people, plus an extra gallon per day because his wife is pregnant. They need half a gallon per day for the animals. Because of the heat in Rob's location, an extra gallon per day is required to ensure adequate supply. He also needs a gallon per day to wash dishes and prepare food. So, the daily drinking water necessary for Rob's household is seven and a half gallons. Multiply this by five days, and Rob should try to store thirty-seven and a half gallons of drinking water before hurricane season.

gallons for 1 day = 4 people + 1 for pregnant wife + 0.5 for pets + 1 for heat + 1 for dishes and food = 7.5

gallons for 5 days = 7.5 x 5 = 37.5

Clare lives alone with her large dog. She has a garden and three laying hens. She would like to be prepared for emergencies by storing thirty days of water for her household. Clare and her dog each need a gallon of drinking water per day. She needs approximately a gallon per week for her chickens to drink. She needs half a gallon of drinking water daily for washing dishes and food preparation, which adds up to seventy-nine gallons for a thirty-day supply. On top of this, she needs non-potable water for other household needs. Her garden is seventy-two square feet, so in summer, it will take thirty-six gallons of water daily for a 30-day total of 1,080 gallons. Clare lives in an area that averages five inches of rain per month in summer. Using a rainfall calculator, she determined that her garden would get 224 gallons supplied by the rainfall during an average summer month. This means an additional 856 gallons of water will be required for her garden during a summer month. She also stores a gallon of non-potable water daily for toilet flushing and non-food household uses and ten gallons per week for bathing. Clare has rainwater barrels with a total storage capacity of 300 gallons. Using her rainwater calculator, she discovers that, based on the average amount of rainfall, she can harvest about 1800 gallons of water from the roof of her 750-square-foot home in a month, assuming the area receives its average amount of rainfall. As long as rainwater fills her barrels three times a month, this will cover her garden's needs. Since they should fill six times during an average month, her non-potable water needs should be covered entirely by her rainwater harvesting system for the month.

gallons for 1 week = 7 for people + 7 for dog + 1 for chickens + 3.5 for dishes and food = 18.5

gallons for 1 day = 18.5 / 7 = 2.6

gallons for 30 days = 2.6 x 30 =78

Location and Placement

The first consideration for emergency water storage is where to keep it. The solutions here differ significantly depending on your living situation and space constraints. Some places to store sealed gallon jugs of drinking water are under sinks in the kitchen and bathrooms, under furniture, in closets throughout your home, in a garage or storage shed, and in the laundry room. (*Emergency Water Storage*, 2020) Unless you want to get a great workout every time you need water, which could actually be a smart exercise strategy, it is important to store as much of your supply as possible in easily accessible areas. Large water tanks should be placed near the house to facilitate filling and use.

Also, remember that water is heavy! Since it weighs approximately eight pounds per gallon, fill large containers where you intend to store them. (*Emergency Water Storage*, 2020)

Take building codes into account when constructing or installing a large water tank. In many states, water tanks can not be installed within ten feet of a property line (Oakville Water Services, 2019). Checking local guidelines will ensure you don't install a water tank you will later be required to move. Water storage tanks come in sizes ranging from 65 gallons to over 1,000 gallons. Knowing how much water you need to store is essential before buying, placing, and filling a water tank.

Long-Term Storage Solutions

Most prepper websites suggest storing water to last at least two weeks, with the standard for preparedness being a month's worth of water. (Brett & Kate McKay, 2021) This can feel like a daunting task, depending on the size of your family and your water needs. Breaking it down into manageable chunks will help you not to get overwhelmed.

The first step is to have small bottles of drinking water, one gallon and under. These are easy to get and portable, essential characteristics if you need to leave your home during an emergency. In countries that regulate bottled water, buying sealed bottles guarantees a sanitary water supply. Although sealed, sanitized water does not spoil, even sealed containers should be replaced after about a year. (Brett & Kate McKay, 2021)

Continue building your water supply using 5-gallon jugs. Label all of your containers with the date they were bought or bottled, as well as whether or not the supply is potable. (Cunningham, 2022)

If your water storage needs are high enough to necessitate larger containers, the next step is a 55-gallon drum. Use a food-safe drum for water storage, even if you don't intend to drink it. Before storing water in a 55-gallon drum, the drum should be washed with soapy water. Make sure you fill the drum with uncontaminated water, seal it tightly, and don't touch the inside of the lid before sealing. (Garcia, 2024)

Ensuring a Drinkable Water Supply

One significant concern of emergency water storage is guaranteeing that the water stored stays usable for drinking. Here are several ways to accomplish this.

- **Use appropriate containers.** If you choose to bottle your drinking water rather than buy it, your containers must be the correct type of plastic or glass, thoroughly cleaned, and securely sealed. Plastic containers that once held milk or juice might seem ideal, but they should not be used for water storage. This is because residue from milk and juice is tough to remove and can lead to water spoilage. (Garcia, 2024) Glass is heavy and breaks easily, but it is an environmentally friendly, easy-to-sanitize option that is easy to sanitize. Assure any glass containers you use are small enough to be transported easily when full and have never held non-food substances. (Cunningham, 2022) Plastic containers are often preferable since they are lighter, non-breakable, and cost less. Safe plastics are marked "food-grade" or labeled with the recycling symbol around numbers 2, 4, or 5. (Garcia, 2024)

- **Clean containers thoroughly.** Plastic containers can be cleaned by thoroughly washing them, inside and out, with soap and water. Then, if a container once held food or drink, it should be sanitized with a solution of 1 teaspoon of bleach per quart of water. Swish this solution around to ensure that it has touched every part of the inside of the container. Heat-safe glass containers can be sanitized by submerging them in boiling water for one minute plus an additional minute per 1,000 feet above sea level. (Garcia, 2024)

- **Fill your containers with safe, drinkable water.**

    - If tap water comes from a reliable water treatment facility, you can use it to fill your containers. If you are filling large drums with a hose, experts recommend using a drinking water hose, which is available from hardware stores. These hoses are made of food-grade plastic, which will ensure that your drinking water doesn't pick up chemicals on the way into your storage container. (Brett & Kate McKay, 2021)

    - If you use water from a spring, stream, rainwater barrel, or any other open source, the water must first be filtered thoroughly to remove contaminants before storing. Pour your water through a mesh strainer to remove any particles. Smaller fabric mesh will remove more particles than wire mesh. If desired, a mesh sock-type strainer can be filled with activated charcoal before the water is poured through it for maximum benefits. Once the water is filtered, it must be treated using one of the methods below.

    - If you store water from any source that has not been treated, including well water, you must treat it before using it for long-term storage.

- **Seal your containers carefully.** Treat the lids of used containers precisely the way you treat the container itself. All the lids you use to seal your water supply should be cleaned and sanitized. Once they have been cleaned, make sure you only touch the outside of the lid when you touch it to avoid any contamination from your hands getting into your water. (Garcia, 2024) Once your containers are sealed, label them with the month and year. The label should also identify whether the water is drinking water or if it is for household use. Open one container at a time so your containers stay sealed as long as possible.

- **Rotate your water annually.** Although water does not spoil, experts recommend using water from containers after a year and refilling them with fresh water. This helps avoid stale-tasting water and ending up with incorrectly bottled water that may sit long enough to develop algae or mold. (Cunningham, 2022)

Your water supply should be stored away from all types of chemicals. Check on it every six months and replace any bottles that have become damaged or cloudy (Garcia, 2024). It is also a good idea to stash drinking water in small, easy-to-carry bottles near an exit in case you need to leave your house during an emergency. When creating your water storage plan, it is wise to store water in containers of various sizes and in different areas of your home.

When I began storing water for long-term emergency use, a friend told me her horror story of using water stored in a tank. The water started smelling bad, and no one knew why. They used it anyway but, after a few days, decided to investigate further. To their disgust, they discovered that a lizard had gotten into the tank somehow, died, and was floating in the water, decomposing. This repulsive experience fueled my determination to vigilantly protect the cleanliness of our stored water so that my family never experiences a similar situation.

Seasonal Considerations

Depending on your geographical area and climate, you may encounter challenges with storing water that vary from season to season. Consider whether any of these situations are likely to create problems for you and plan accordingly.

Winter Considerations

The biggest problem for storing water over the winter months is freezing. When water freezes, it becomes unusable because it cannot be removed from whatever container it is stored in, which would be problematic if an emergency occurs in winter. Water also expands by approximately 10% when it freezes. (Walter, n.d.) This presents a hazard, as pipes, tanks, and bottles may crack when water freezes. Unnoticed cracks can result in water contamination or leaking. There are several ways that you can protect your water from freezing.

- **Store as much of your water as possible inside your home.** While many households will also need outdoor storage, storing as many of your containers as possible indoors will ensure that they remain above 32°F to avoid freezing. It is especially important to keep small containers indoors since they will freeze much more quickly than larger water tanks.

- **Insulate water that is kept outside.** Tank covers can be bought to protect your water tank from freezing. Foam insulation is another option, but choose insulation made specifically for water tanks if you go this route. Since the foam is injected into the inside of the tank, it must be designed to avoid contamination if the foam breaks down. (Blair, 2023) Reflective covers are a good option, as they are designed to reflect the heat from the water back into the tank. Whatever type of insulation you use, the tank should be covered on all sides except the bottom. This is because the tank will absorb the warmth from the ground or the air rising from below. (Tanks for Everything, 2022)

- **Store your water supply in a greenhouse.** Because the temperature in greenhouses is higher than outside, they make ideal solutions for winter water storage. Greenhouse temperatures can be further raised by adding an open compost bin. (Walter, n.d.)

- **Protect your water tank from the wind.** Suppose you can't place your tank in a sheltered location. In that case, you can use a wind deflector or tarp to create a windscreen. This will prevent freezing winds from accelerating the temperature drop in your tank during a winter storm.

- **Use one large, round tank instead of several smaller ones.** While smaller tanks may be easier to handle and fill, they are more prone to freezing because they have more surface area. If you are in a freeze-prone location, using one larger tank rather than a few smaller tanks will decrease the likelihood that your water will freeze. (Walter, n.d.) Also, round tanks have less surface area for their mass, so a tank with rounded sides rather than a cubical tank is less likely to freeze during a harsh winter. (Tanks for everything, 2022)

- **Position your outdoor tank in a sunny spot.** Tank placement is important when trying to prevent freezing. A tank that receives direct sunlight several hours a day will absorb warmth that will help keep it from freezing overnight when the temperatures are at their lowest. In the northern hemisphere, this means a south placement is usually best. In the southern hemisphere, place your tank on the north side of your house. Tanks can be painted black for added sun absorption. (Walter, n.d.)

- **Use a tank heater or de-icer.** Tank heaters come in a variety of sizes and price points. Be careful to buy a tank heater that is the right size for your tank. You should also consider your tank's materials, as some tank heaters are unsafe to use in a plastic tank. (Blair, 2023) A variation on a tank heater is a de-icer. While these don't change the temperature of the water, they use an electric current to prevent ice from forming in the tank. These options work if you have access to a power source, whether electric, propane, or solar.

- **Keep the water moving.** Another way to reduce ice formation is to keep water moving, which a bubbler does. Alternatively, use and replace water in your tank daily during the winter. If your outdoor tank is small enough, simply rocking it around during a freeze may be enough to keep the water in it from freezing. (Tanks for Everything, 2022)

Summer Considerations

Keeping water from going bad during the summer is much easier if stored correctly from the start. Water should be kept in cool, dark locations. During warmer months, don't store water where it will be in direct sunlight or get excessively hot. (*Emergency Water Storage*, 2020)

Creating Your Emergency Water Storage Plan

As you can see, figuring out how much water to store for an emergency is complex. Use the steps below to decide how much water you need to store to meet your water emergency goals.

- How many people are you storing water for?
- Are you storing water for pregnant women or people with special needs?
- Do you have pets or livestock to provide for?
- Do you need water for irrigation?
- How many days of emergency water use would you like to be prepared for?
- Have you included water for cleaning, bathing, food preparation, and toilets?
- Use a rainfall calculator like the one found at https://water.usgs.gov/edu/activity-howmuchrain.html to ascertain how much rain you get on average.
- Use the difference between how much water you need in an emergency and how much rainfall you get on average to decide how much water you should store for emergency use.

Strategy Matters

During my time on the Amazon, we routinely filled pots and pitchers with water in the morning in case the community water tank ran dry in the afternoon. This didn't often happen, but it could be problematic if it caught you unprepared when you wanted to cook dinner, make coffee, or wash your hands. Filling a few pitchers ahead of time was a simple way to ensure that your afternoon could proceed smoothly if there were a few hours without water.

Similarly, I have simplified the process of creating a water storage strategy to ensure your life can proceed smoothly even during a water shortage emergency. Learning to store water safely and protect it from contamination, leaks, and freezing will help safeguard your carefully acquired water supply. In the next chapter, we will delve more deeply into different purification techniques to make water safe for drinking.

4

Simple Water Purification Techniques

> "Water does not resist. Water flows. When you plunge your hand into it, all you feel is a caress. Water is not a solid wall, it will not stop you. But water always goes where it wants to go, and nothing in the end can stand against it. Water is patient. Dripping water wears away a stone. Remember that, my child. Remember you are half water. If you can't go through an obstacle, go around it. Water does." — *Margaret Atwood*

Harvesting and storing water can ensure a water supply during an emergency. However, without knowledge of purifying techniques, you might still be left with undrinkable water. This makes learning to purify water from any source essential to your water preparedness strategy.

Boiling

Boiling water has been the most widely used method of ensuring water safety for centuries. It is also the earliest known system of disinfecting water, with the first mention of boiling water to purify it dating from about 2000 BC. (ZeroB, 2024)

Advantages of Boiling

Boiling is a highly effective method of killing or deactivating all bacteria, viruses, and protozoa commonly found in untreated water. These pathogens cause gastrointestinal illness.(Shrestha et al., n.d.) Boiling water before drinking is acknowledged to be the only simple purification technique that is 100% effective in killing all of the following disease-causing pathogens. (The Survival University, 2024)

- Cryptosporidium
- Giardia
- Campylobacter
- Salmonella
- Shigella
- E. coli
- Enterovirus
- Hepatitis A
- Norovirus
- Rotavirus

Disadvantages of Boiling

While boiling is the most effective method of clearing water of disease-causing pathogens, some impurities are not removed by boiling water. Boiling doesn't remove any chemical pollution or metals. (Shrestha et al., n.d.) Boiled water can also have a flat taste that many people dislike, although this can be remedied by adding a pinch of salt. (CDC, 2020) Also, water is lost as steam during boiling, which can be a problem in a water shortage.

Another disadvantage of boiling water is that it requires fuel. In an emergency where fuel is scarce, boiling may not be a feasible purification method.

How to Boil Water for Sanitization

- If necessary, filter your water before boiling to remove debris.
- To be considered sanitized, water must be brought to a rolling boil for a full minute or three minutes at 5,000 feet or higher above sea level. (Garcia, 2024)
- After removing the water from the heat, add a pinch of salt to each quart or liter to normalize the taste.
- Cover and allow the water to cool to room temperature.
- Next, pour it into pre-sanitized jugs and seal them until ready to use them.

Solar Disinfection

Solar disinfection, commonly known as SODIS, is considered one of the most useful and cost-effective treatment options for individuals purifying water. (García-Gil et al., 2021) SODIS combines heat and UV rays to kill harmful pathogens in untreated water.

How SODIS Works

SODIS uses a mixture of UV rays and heat to sanitize small quantities of water in clear bottles. UV rays effectively kill disease-causing bacteria, viruses, and protozoa that can survive in the gastrointestinal tract. (UN Environment Programme, 2016) Heating water in direct sunlight utilizes UV rays to kill or deactivate biological impurities in untreated water. SODIS is as effective as boiling for killing pathogens as long as it is done correctly. However, it has a higher rate of uncertainty due to variables in weather conditions. (García-Gil et al., 2021)

Advantages of SODIS

The most significant advantage of SODIS is that it uses free energy from the sun to make untreated water safe to drink. This makes it a viable option in situations with limited fuel availability. It is estimated that sanitizing water by boiling can cost up to $10.56 per person per year while using SODIS to sanitize water costs about $0.63 per person per year. (García-Gil et al., 2021)

SODIS kills the same pathogens as boiling and carries less risk of recontamination because the water can be sanitized in the same bottle it will be stored in (Garcia-Gil et al., 2021). Since individuals reuse plastic drink bottles to sanitize and store clean water, SODIS is an environmentally friendly option. The most common plastic bottles in use are PET plastic bottles, and tests have shown that PET plastics do not leach chemicals into water when heated. (Regula Meierhofer, 2002)

Disadvantages of SODIS

SODIS is not a quick method of water decontamination. It will take six to forty-eight hours to purify untreated water, depending on the size of the bottles and the amount of direct sunlight they receive. It is recommended that individuals use 2-liter or smaller bottles for SODIS. (García-Gil et al., 2021)

Water that is not clear is not suitable for SODIS. Suspended sediment in the water blocks UV rays from penetrating the water to destroy the pathogens. (UN Environment Programme, 2016) Additionally, SODIS doesn't remove chemicals or metals from water. It also cannot be used on days of continuous rainfall. (Regula Meierhofer, 2002)

Climate Zone Considerations

The further you are from the equator, the more difficult it becomes to use SODIS as a water purification method. (Regula Meierhofer, 2002) This is because the water will not reach a sufficient temperature for sanitization, so the UV rays must work alone to kill pathogens. Studies have shown SODIS to be highly effective in tropical and subtropical climates, up to 34° away from the equator. This means SODIS is viable in cities as far south as Sydney, Australia, and as far north as Los Angeles, California. (Davies et al., 2009) However, if you are further from the equator than this, in a temperate, sub-polar, or polar climate, you might still be able to use SODIS.

There are two types of UV rays, UVA and UVB. Without the added benefit of heat that comes with being closer to the equator, you need to get both types of UV rays working for you. Unfortunately, PET plastic bottles will stop most UVB rays from reaching the water, making SODIS less effective in colder climates. (García-Gil et al., 2021) Studies conducted in Finland have shown that water can achieve similar levels of decontamination in colder climates if PE plastic bags are used for SODIS instead of PET bottles. This is because the PE plastic doesn't block the UVB rays. (Juvakoski et al., 2022) PE bags are available in most places as Ziploc-style storage and freezer bags.

How to Sanitize Water Using SODIS

- If necessary, filter your water to remove debris and increase the clarity.
- Wash clear plastic drink bottles or glass bottles thoroughly before filling them.
- Fill the bottles with the clearest water available.
- Lay the bottles on their side in full sunlight. A piece of corrugated tin is perfect for holding bottles. Placing bottles on a black surface increases the efficiency of SODIS.
- Expose bottles to full, bright sunlight for six continuous hours, increasing the time if weather conditions are not ideal. In partly cloudy conditions, bottles should be exposed to the sun for 48 hours.

- Water purified using SODIS should be consumed within two days or kept cold to avoid bacterial regrowth. (Rosa e Silva et al., 2021)

Simple Filtration

Simple filtration systems can be made for under $15. If your water supply is cloudy or has particles floating in it, filtering is recommended before sanitizing. (Shallcross, 2023) Most filtration systems use a combination of elements for filtering since different components will remove specific contaminants.

Advantages of Handmade Filters

Making a water filter allows you to choose its components to filter the contaminants most likely in your water supply. The biggest advantage of homemade filters is that the proper filtration process will enable you to turn practically any available water into drinking water. Some filters are reusable, only needing to be replaced after filtering many gallons of water. Handmade filters are economical and practical solutions for providing clean water in an emergency.

Disadvantages of Handmade Filters

Handmade filters are small and slow. Water may take hours to drip through to the bottom. While the positive side is that a slower filtration process ensures cleaner water, the negative side is waiting for the water to be ready for drinking. Also, you may lack the proper filtering components. In that case, microscopic pathogens could still be present, necessitating a second purification method to ensure a drinkable water supply.

Creating a Filter

Creating a handmade filter is an easy and economical project. You will need to gather the following supplies.

- Craft knife or scissors

- Clean plastic 2-liter bottle with a cap
- String (optional)
- Hammer and nail
- Coffee filter, paper towel, or cheesecloth
- Large cup
- Container to catch the filtered water
- Banana peel - filters bacteria and metals (Rosario et al., 2022)
- A thin stick cut from a pine tree, pre-soaked in water for an hour - filters 99% of pathogens (Chu, 2021)
- Cotton balls - to remove any charcoal residue (Michaud, n.d.)
- Activated charcoal - filters organic compounds and chemicals such as chlorine (Taraba & Heaton, 1990)
- Clean sand - filters suspended matter (EMIS, 2010)
- Gravel or small rocks - filters sediment and larger suspended solids (BMS, 2023)

Make the filter according to the following directions. Since each component filters different contaminants, any missing filtration element reduces the quality of the filtered water. Filters that do not contain banana peel or pine sticks do not filter out pathogens, meaning that if you do not have these materials for your filter, water should be sanitized by boiling or SODIS before drinking.

1. Cut the bottom off of the plastic bottle using your scissors or knife.
2. If you want to hang the filter while it filters the water, use the nail to poke two holes opposite each other about an inch from the cut edge. Securely tie a piece of string in both holes to make a handle from which it can hang.
3. Remove the cap from the bottle, then use the hammer and nail to drive one or

two holes into the top of the cap. More holes would result in a faster filtration time and lessen the effectiveness of your filter, so do not put more than 2 holes in the cap.

4. Place the coffee filter, paper towel, or cheesecloth on the mouth of the bottle, then screw the cap over it.

5. Place the bottle, mouth down, into a large cup to help hold it steady while filling it.

6. Place a few cotton balls into the neck of the bottle.

7. Finely chop the banana peel and layer it on the cotton balls.

8. Using a utility knife, remove the bark from the pine stick. Break or cut the stick into pieces that fit in the bottle and place them on top of the banana peel.

9. Fill a third of the remaining space in the bottle with activated charcoal. Sometimes called activated carbon, this can be purchased in small pieces or powder form.

10. Add sand on top of the charcoal. If you have fine and coarse sand, use both, as each works differently. Place the fine sand on top of the charcoal and the coarse sand above that.

11. Fill the rest of the bottle with gravel, leaving about an inch at the top empty.

12. Place the filter, cap side down, into the mouth of a container. If you added a handle in Step 2, you can hang it above the container.

Because this filter contains biological components, it is not reusable. A reusable alternative would be to assemble the filter without the banana peel and pine sticks. This would allow you to reuse the filter many times before replacing it. Then, assemble a separate filter in a smaller bottle with cotton balls at the bottom, banana peels, and pine sticks on top. You should only make one hole in this bottle's cap to slow the water flow through the filter.

In this option, you first filter the water through the large filter, then follow that up by filtering it through the smaller one. This way, you can disassemble only the biological filter. After you remove them from the filter, the banana peels can be used again as long as they are allowed to dry thoroughly between uses. They can be used up to ten times before losing their filtration capabilities. (Rosario et al., 2022) The pine stick should be replaced each time you use the filter. (Chu, 2021)

Using Your Handmade Filter

Ensure that the container your filter is sitting in is large enough to hold the water you plan on filtering. Slowly pour water into the top of the filter until the gravel is covered. As the water level goes down, add water to cover the gravel again until all the water is filtered.

Wait until all the water has dripped through the filter before removing the container from the bottom.

If you have a second biological filter, you should repeat the pouring process over the second filter, covering the pine sticks by pouring slowly and then allowing the water to pour out of the bottom into another container. This container should be closed tightly once full and stored cold or at room temperature. Do not open it until you're ready to drink it.

Chemical Purification

Chemicals are commonly used to disinfect drinking water in municipal water systems. City water is often treated with a several-step process that includes screening, chemical coagulation, flocculation, sedimentation, clarification, filtration, and disinfection (Hancock, 2016). Each step removes specific impurities.

Screening removes the large debris. Next, the addition of coagulating chemicals increases the particle size of impurities. Flocculation is the gentle agitation of the water, which encourages the larger particles to bind together. During sedimentation, the water sits for several hours, allowing the particles to sink to the bottom of the tank. Next, the water is drained from the top and filtered through sand, gravel, and activated charcoal to remove suspended matter, organic compounds, and chemicals. Lastly, the water is treated with chemicals to kill the remaining bacteria and pathogens. (Stricklin, 2022)

Chemical purification can be adapted for household use to ensure safe drinking water. Several chemicals are commonly used to address water safety concerns, and each works differently to kill or inactivate pathogens.

Chlorination

Chlorination is a well-known way to sanitize your water for long-term storage. Chlorine was first added to water in Britain in the early 19th century to remove the odor. Discovering that water was spreading disease due to pathogens that were present, scientists suddenly realized chlorine could be an effective tool for disinfecting water to eliminate water-borne illness. (Hancock, 2016) Even today, chlorine is acknowledged to be one of the most effective water-disinfecting chemicals.

Chlorine kills or inactivates pathogens by attacking the cell wall. Once the cell wall is weakened, the chlorine can disrupt respiration and the DNA replication of the cells. (Hancock, 2016) You can sanitize water by adding 1/8 teaspoon of unscented household bleach per gallon. Use 1/4 teaspoon of bleach if the water you're treating is cloudy. Allow the water to sit for 15 minutes before drinking or bottling for long-term storage. If treated water still appears cloudy or doesn't have a slight chlorine smell, treat it again. (Brett & Kate McKay, 2021) Any leftover chlorine taste should be gone once the water sits for several hours, even if it is still present when you bottle it. (EPA, 2015)

While chlorination is effective against almost all disease-causing pathogens, the protozoa Giardia and Cryptosporidium are two exceptions not destroyed by chlorine (Hancock, 2016). If you are concerned that your water may contain either of these pathogens, use another sanitization method, such as boiling or SODIS.

Iodine

Iodine is common in first aid kits and medical supplies because of its disinfecting qualities, which can also sanitize water for drinking. Iodine, similar to chlorine, kills or inactivates bacteria by penetrating the cell wall and disrupting the function of the cell. While it was over a hundred years ago that scientists figured out the mechanism chlorine uses, how iodine works to kill pathogens is still something of a mystery. We do know that it causes such rapid oxidation of different cell parts that it overwhelms the pathogen's ability to repair itself. (Eggers, 2019)

Liquid iodine (2%), generally used for medicinal purposes, can be used to disinfect water for drinking by adding twenty drops per gallon of water. Pregnant women, people who have a thyroid issue, or those with a known iodine sensitivity should not drink water treated with iodine. It should also not be used for more than three weeks out of any twelve months to avoid potential thyroid damage. (Environmental Public Health, Alberta Health Services, 2021)

Neither iodine nor chlorine is effective against the parasite Cryptosporidium. If there is a risk that Cryptosporidium is present, combining iodine drops with SODIS is believed to be an effective way to ensure the water is safe to drink. (Environmental Public Health, Alberta Health Services, 2021)

Water Tablets

Water purification tablets are available at many camping or sporting goods stores, pharmacies, and online. These tablets contain chlorine, iodine, other disinfecting chemicals, or a mixture. Different types of tablets differ in strength and effectiveness, so be sure to follow the package directions for your tablets, as these directions are designed for that specific combination of chemicals. (EPA, 2015)

Indicator Tests

Many types of water treatments purify limited contaminants out of your water supply. Testing your water is a logical first step to knowing what purification techniques to employ. This knowledge can save you money on unnecessary treatments and ensure you choose the appropriate methods to remove contaminants specific to your water supply.

Types of Tests

The three main types of water quality testing are bacterial, mineral, and pH tests. To be considered safe for drinking, water should be free of disease-causing pathogens, have low concentrations of toxic metals or chemicals, and be clear and tasteless. (Auriga Research,

2021) While you can hopefully figure out on your own whether your water is clear and tasteless, pathogens and other pollutants can go undetected without proper testing.

Test Strips

Since test strips rely on user interpretation and cannot show precise contaminant levels, they are less accurate than other tests. Because different users interpret the colors on the strips differently, researcher Lloyd Wilson says, "If you give a strip to five different people, you might get five different results." (Sloan, 2021) However, they are helpful for an initial assessment of the safety and quality of your drinking water.

Test strips are not regulated by any central agency, so purchase them from a reputable company with positive reviews. Comprehensive test strips can be found online to test for coliform bacteria, lead, nitrites, pesticides, and chlorine and supply information regarding the water's pH levels. At the time of writing, they can be purchased for under $50.

Color Disk Kits

Color disk kits use the same concept as test strips but with a few added steps, which allows for a more accurate reading.

1. Drip water from a sterilized dropper into the reagent.
2. The reagent reacts with whichever contaminant is being measured, causing it to develop a specific color.
3. Wait for the amount of time specified on your kit.
4. Insert the tube of reagent into a viewing box that contains a gradient.
5. Match the color of the reagent to the closest color on the gradient. (Lawson & Mistry, 2017)

Color disk kits have more steps and are typically more complex than test strips. This can lead to a higher chance of user error. However, if used correctly, these tests can give a more accurate picture of the contaminants in your water. Due to their complex nature, they usually focus on fewer contaminants, so one kit may test for chlorine, while a separate test would need to be purchased if you want to test for PFAs. Each kit costs upwards of $100 at the time of writing.

Digital Instruments

There are two types of digital water testers. One type costs about a thousand dollars, needs calibration, and is designed for use in a lab. The other type can be purchased online, costs less than $25 at the time of writing, and is intended for portable use. These testers only provide general information about whether water is safe to drink. They don't tell you why the water is unsafe. These small, economical instruments would be useful if tap water is still available but may be contaminated.

Laboratory Testing

If you dig a new well or are concerned about the cleanliness of the water currently coming from your tap, the EPA recommends sending it to a lab for testing. This is the most accurate way to know what contaminants may be present. (Environmental Protection Agency, 2018) Laboratory testing takes one to two weeks since the water must be sent to an EPA-approved lab. It can also cost hundreds of dollars. However, testing can be worth the wait and expense if you have tapwater quality concerns.

Simple Water Purification Chart

Boiling	SODIS	Filtration	Chemical Purification
Kills 100% of pathogens	Kills 99% of pathogens if weather conditions are conducive	Kills 99% of pathogens if the correct materials are available	Kills all pathogens, except Giardia and Cryptosporidium
Does not remove heavy metals or toxic chemicals	Does not remove heavy metals or toxic chemicals	Removes sediment, most metals, and toxic chemicals	Can improve the taste and smell of untreated water
Heavy fuel consumption	Low cost due to no fuel consumption and reusable supplies	Filter components must be available for the filter to be effective	Cost-effective and convenient
Low level of user error	Water takes hours or days to purify	Slower filtration times generally result in cleaner water	Iodine can result in thyroid problems if used over a prolonged period

Teamwork Wins

The most pressing concern in a short-term water emergency is that water does not have bacteria or other disease-causing pathogens. In this situation, boiling water is the most effective disinfecting solution. However, it is an insufficient long-term solution because boiling water does not remove toxic chemicals or heavy metals. Stacking purification methods into "teams" of purifying strategies is preferable for a long-term water crisis. So filtering water through activated charcoal, sand, and gravel, followed by boiling, is a great purification team. Similarly, chlorination and SODIS can work together to ensure water is suitable for drinking.

These simple purification techniques each have strengths and weaknesses. They are strategies for when you find yourself in a water emergency. In the next chapter, we will get into more advanced purification techniques that can be set up ahead of an emergency to provide peace of mind and ensure that your family is ready beforehand if an emergency arises.

Before we dive into these more advanced purification techniques, if you are gaining value from this book and want to pass along the goodness and water survival secrets to others, would you please take 2 minutes and leave a quick review?

Here is the Amazon link and a QR code:

Now let's get back into our journey of learning and implementing more water survival secrets!

5

Advanced Purification Systems

"The earth, the air, the land, and the water are not an inheritance from our forefathers but on loan from our children. So we have to hand over to them at least as it was handed over to us." — *Mahatma Gandhi*

ADVANCED WATER PURIFICATION SYSTEMS commonly use a multi-barrier approach to water purification. These systems can be used in your home, providing a way for you to ensure a clean water supply today as well as proactively prepare for an emergency. Homes with whole-house purification systems can continue to function without a hitch if there is a boil warning from the local water company. Even a small countertop filtration system can enable you to turn rainwater into drinking water if necessary.

Types of Advanced Purification Systems

Different types of systems can be installed to purify all the water that comes into your home. Smaller units can be bought to filter the water from a specific area, such as the kitchen sink. Installing an advanced system before a water emergency occurs ensures you are prepared for whatever happens. There are several advanced systems to choose from, depending on your budget and what contaminants need to be cleaned from your water supply.

Many advanced purification systems come in different varieties. Systems that treat water as it enters your home are called point-of-entry units. These units provide treatment in one area, usually a garage, for all water that comes into your home. Point-of-use units are smaller. They may attach to a faucet or showerhead or sit in the cabinet under your sink. Either way, point-of-use units treat the water coming out of a specific outlet in your home.

Commercial Water Filters

In about 500 BC, Hippocrates invented what is believed to be the first household water filter, appropriately dubbed The Hippocratic Sleeve. Boiled water was poured through the cloth bag to remove impurities and trapped sediment. (Water Filters Australia, 2017) Water filtration has advanced considerably since then, with new materials to filter through, but the basic principles have remained the same.

Activated Carbon Filters

These filters use activated charcoal to purify water, similar to filters made at home. The carbon is activated by being treated with oxygen, carbon dioxide, and steam to create more pores, increasing the internal space for the carbon to trap microscopic contaminants from the water. (Patel, 2018) Activated carbon has been used since the early nineteenth century to treat poison overdoses, and its use to filter out toxins from water began much earlier. Thousands of years ago, Phoenicians and ancient Hindus wrote about filtering water through charcoal. (Sappo Hill Soapworks, n.d.)

Activated carbon filters help filter out or reduce many common contaminants. This includes chlorine, arsenic, pesticides, herbicides, nitrates, PFAs and PFOs, lithium, and microplastics. They cannot be relied on to rid water of bacteria, microbes, or other natural pathogens. (Franks, n.d.) Carbon filters range in size from pitchers on your kitchen countertop to full-house systems requiring professional installation. Carbon filters must be replaced according to the manufacturer's directions, since water quality may actually worsen after being filtered through a worn-out carbon filter. This is because stored toxins may break off the activated carbon while it is trying to absorb new ones. (Woodard, 2019)

Carbonless Filters

Chlorine Injector

One type of carbonless filter, commonly known as a chlorine injector, uses chlorine and monochloramine instead of activated carbon to filter impurities out of water. Monochloramine is a chlorine derivative often used to treat water because it is weaker and doesn't affect the water's taste and smell like chlorine can. It is also less likely to accumulate to unsafe levels in water. These can be used to rid your water of bacteria and other natural pathogens. (*How Does a Commercial Water Filtration System Work?*, 2024) However, they are less effective than activated carbon filters at removing toxic chemicals and microplastics. (Franks, n.d.)

Chlorine injectors are commonly sold to treat well water because almost all municipal water is treated with chlorine. Some chlorine injectors also contain a charcoal filter to remove the chlorine and other chemicals before the water enters your home. This double-filtration provides all the benefits of a chlorine injector plus the benefits of a carbon filter, making it a highly effective method of ensuring access to clean water.

Ceramic Filter

Another type of carbonless filter is a ceramic filter. These filters have been used since the late nineteenth century, when a British potter named Henry Doulton began making them for Londoners who gathered drinking water from the cholera and typhoid-infested Thames (Kiger, 2013). Ceramic filters are made from clay mixed with sawdust to create pores large enough for water molecules to seep through but not large enough for bacteria and other contaminants to pass through.

Ceramic filters are effective against most bacteria and protozoa, but viruses can usually avoid being trapped due to their smaller size. (CDC, 2008) These systems are often made more effective by adding activated charcoal or silver to the clay, which traps and kills pathogens as they try to make their way through the pores. (Kiger, 2013) They do not remove most toxic chemicals or metals but are effective against most microplastics. (*How to Filter & Remove Microplastics from Tap Water*, 2018)

Ceramic filtration systems are generally small, ranging from personal water bottles to tabletop jugs that hold up to six liters and use gravity to push water through the ceramic filter. The most common ceramic filter for emergency use is called a candle filter. These can be used with a pump or siphon to push water through the filter into a clean container much more quickly than a gravity filter.

UV Light Sterilization

UV light sterilization works on the same principles as solar disinfection (SODIS), which we discussed in the previous chapter. Water is passed through a UV light during filtration to purify it. This process is entirely chemical-free but kills or inactivates 99.99% of disease-causing pathogens. (Royer, n.d.)

Water must be clear for UV light to sanitize it. Murky water or water with dissolved solids must be filtered for clarity for UV light filtering to work. Many UV filters include a pre-filter to increase water clarity and enhance effectiveness. UV light cannot filter out minerals or toxic chemicals, although it is highly effective against bacteria, viruses, and protozoa. (Hague Water Supply, 2019) You'll need to install a backup power source for your UV filter since any interruption in power during UV filtration will result in water that is not properly sanitized. (Royer, n.d.)

Distillation Techniques

Distillation is the process of boiling water until it evaporates and condenses. First, the water is boiled in an enclosed chamber until it evaporates. Most contaminants cannot evaporate, so the water is cleaned of these impurities. The steam is then passed through a cooling chamber, where it condenses back into water droplets. These newly-purified droplets drip into a clean container. In most water distillers, once the water has condensed it is run through an activated charcoal filter to remove any impurities that were able to evaporate and condense with the water. This is a slow process, with each gallon of distilled water taking up to six hours to produce. (Campbell, 2022)

Distilled water is considered the purest water available because the distillation and filtration process can desalinate water and remove 99.9% of impurities. (Valley Food Storage, 2023) Distillation removes heavy metals, dissolved solids, and toxic chemicals such as nitrate, fluoride, sodium, and sulfate. It also kills most bacteria, viruses, and pathogens. (*Publication No. 7: Distillation Treatment Systems for Private Drinking Water System*, 2009) A distiller with an activated charcoal filter filters out the volatile organic compounds (VOCs) that distillation leaves behind. (Environmental Protection Agency, 2019)

Distillation systems are available as small countertop units that can distill one to three gallons of water at a time. They are also available as full-house units that must be professionally installed. Smaller units use about the same amount of electricity as a microwave, while larger ones use significantly more. (*Do Water Distillers Use a Lot of Electricity?*, 2024)

Reverse Osmosis

Reverse osmosis uses pressure to reverse the natural process of osmosis. In osmosis, water moves naturally through a semi-permeable barrier from an area of more water to one with less. The barrier stops dissolved particles, so the water that moves is naturally "filtered." (Lopez & Hall, 2023) Reverse osmosis uses applied pressure to force water through one or more semi-permeable barriers to achieve the same result as natural osmosis in a controlled way. (Kershner, 2008)

Reverse osmosis can be used to desalinate water, meaning this process can make drinking water from salty ocean water. It can also remove or reduce many minerals and toxic chemicals. (Kershner, 2008) This includes sodium, chloride, chromium, copper, arsenic, radium, sulfate, fluoride, calcium, magnesium, lead, potassium, nitrate, and phosphorous. It is also effective against bacteria, viruses, and protozoa. (CDC, 2020b)

While reverse osmosis filtration is a superior solution to many water quality problems, it has a few disadvantages. The first is that the filter easily clogs because of the microscopic pore size. Second, reverse osmosis wastes significant water, making it a poor choice in a water shortage. The third disadvantage is that it removes everything from the water, even the healthy minerals, which can alter the water's taste and eventually lead to mineral deficiencies for people drinking the filtered water. (Crystal Clear Water, 2022)

Advanced Water Purification Chart

	Activated Carbon Filters	Chlorine Injectors	Ceramic Filters	UV Light Filters	Distillation Systems	Reverse Osmosis System
Effective Against	Toxic chemicals Microplastics & PFA/PFOs Volatile Organic Compounds (VOCs)	Bacteria Other pathogens	Bacteria Protozoa Microplastics	Bacteria Other pathogens	Bacteria Other pathogens Toxic chemicals Microplastics, PFA/PFOs Used for desalination	Toxic chemicals microplastics, PFA/PFOs Bacteria Other pathogens Used for desalination
Not Effective Against	Bacteria Other pathogens	Toxic chemicals Microplastics PFA/PFOs VOCs	Viruses Toxic chemicals Heavy metals VOCs	Toxic chemicals Microplastics PFA/PFOs VOCs	VOCs	VOCs
Cost at Time of Writing	Pitcher - about $20 Point-of-use unit - $150-300 Point-of-entry unit - $1,000 - $2,500	Point-of-entry unit $500-$1,500	$90-$250, depending on the size of the unit	$300-$3,000, depending on the size of the unit	1-gallon container - $100 Point-of-use unit $800-$2,000 Point-of-entry unit- $8,000 $10,000	Point-of-use unit - $150-$300 Point-of-entry unit - $2,000-$4,800
Energy Use	None	Low or none	none	Low, must have continuous power	Medium to high	Medium
Required Care	Replace the filter in a smaller system every 6-12 months, in a full-house unit every 2 years.	Chlorine tanks must be refilled as needed. Check valves and pumps every 6 months and replace as needed.	Wash the ceramic filter regularly to avoid bacterial growth in the filter.	Change UV bulb every 9-12 months Clean the UV sleeve every 6 months and replace every 2 years. Check filters and O-rings and change them as needed.	Clean small systems after each use & clean monthly with vinegar. Follow manufacturer directions for large systems. Replace activated carbon filters as needed.	Replace filters every 6-12 months. Replace the reverse osmosis membrane every 3-5 years.

Think About It

In light of the advanced water purification techniques discussed in this chapter, what do you believe are the most pressing challenges or opportunities in ensuring access to clean and safe drinking water in an emergency?

How can you use these techniques to help address these challenges for yourself, your family, and your community?

Do You Just Want to Survive?

This chapter shows that many intricate systems can be used to purify water. Safe drinking water is non-negotiable for survival, and you are now prepared to harvest, store, and purify water to survive a water emergency. But survival isn't our only goal! We want to thrive. To do that, we must learn adaptability and sustainable habits. These skills are an integral part of the prepper mindset. In Chapter 6, we will discuss practical ways to thrive during an emergency situation and how to make sustainability part of your everyday life today. That way, in an emergency, you will be prepared to make the most of every drop of water you acquire.

Free Bonus Gift

As you pursue a sustainable lifestyle, are you ready to unlock the secrets of how to identify and harvest four common plants that you can use for food in case of emergency, or anytime? These plants might even be growing in your backyard!

Foraging is the practice of seeking and collecting wild foods for sustenance, natural medicine, and more. **Foraging provides the perfect excuse to slow down from the hectic pace of modern life.** Enjoying the beauty of nature while spending time with friends and family is a winning combination that contributes to healthy relationships, souls, minds and bodies. And on top of that is the satisfaction and joy found in sharing a foraged meal or snack together.

In case you're new to foraging, this quickstart foraging guide includes some **tips on safety and getting started**. Then comes detailed **information on each of four common herbs** so you can confidently and successfully **find, identify, harvest, and enjoy** them. General cooking suggestions and **two delicious recipes** are included for each plant.

Backyard Herbalism, your quickstart foraging guide, is ready for you to download by clicking on the following link (if reading the eBook) https://bit.ly/backyardherbalism

or scanning the QR code with your phone (if reading the print version).

6

Thriving, Not Just Surviving

"I love to watch the movement of light on water, and I love to play in rivers and lakes, swimming or canoeing. I am fascinated by people who work with water - fishermen, boatmen - and by a way of life that is dominated by water." — *Berlie Doherty*

In Chapter 1, we talked about having a prepper mindset. If you remember, the three main components of that mindset are sustainability, resourcefulness, and emergency preparedness. All these characteristics are necessary to survive and thrive in an emergency. Let's explore some sustainability habits first.

Sustainability: Water Conservation Habits

Sustainable living means living in a way that the earth can continue to support its inhabitants indefinitely. Humanity is currently on a collision course with a water deficit. (Klobucista & Robinson, 2023) Using sustainable practices to reduce your water consumption now, before a crisis develops, enables you to be a part of the solution.

Reduce Your Use...

In the Bathroom

- **Shorten your shower** - Showers can be a nice break from life, but relaxing under the hot spray comes at an environmental cost. Older showers can use five to ten gallons a minute. (Spain, 2016) A shorter shower lowers your usage. Even better, installing a reduced-flow showerhead can limit water consumption to less than two gallons per minute when showering. (Environmental Protection Agency, 2016)
- **Take baths** - Taking baths instead of showers is a great way to reduce water usage. Fill the tub halfway, which uses less water than even short showers. However, if you have a large garden tub or jacuzzi in your bathroom, this may not save water.
- **Place a bottle in your toilet tank** - Take an empty plastic drinking bottle and fill it with small rocks or a few inches of sand. Screw the lid on tightly. Placing one of these bottles in the tank of each toilet in your house can save as much as ten gallons of water per day. (Eartheasy, 2018) This is effective because it reduces the amount of water each flush uses; the bottle's volume reduces the tank's water capacity.
- **Turn off the water** - The water doesn't need to run while brushing your teeth, shaving, or lathering your hands. You can also install a faucet aerator, which reduces the amount of water coming from the faucet without reducing perceived water pressure.
- **Check your toilet for leaks** - Even small leaks can add up, wasting significant amounts of water. Dozens of gallons per day can be wasted simply by water leaking from your toilet tank into the bowl. You can check for a tank leak by adding some food coloring to your tank. If the water in your toilet bowl begins

to change color without flushing the toilet, call a plumber or replace your toilet tank gasket. (Spain, 2016)

In the Kitchen

- **Use the dishwasher** - A newer dishwasher uses between three and four gallons of water to do a load of dishes, less than washing dishes by hand. (Williams, 2023) Avoid running it until it's full for maximum water savings. If dishes usually sit for a few days before the dishwasher fills up, running the rinse cycle will help prevent dishes from getting smelly while they sit but uses less water than rinsing by hand.

- **Wash vegetables in a basin**—Rather than running water over your vegetables while cleaning them, fill a basin with water. Place vegetables into the basin to rinse. Clean them with your veggie brush, give them a final swirl around in the water, and they are ready to cook.

- **Reduce food waste** - While reducing food waste may appear unrelated to water consumption, it can play a huge part. Agriculture accounts for 70% of the water used each year, and nearly half of the food produced in America is wasted yearly. (Cho, 2011) An average American throws away over three hundred pounds of food per year. (*Food Waste in America in 2020: Statistics & Facts*, n.d.) When you consider that it takes 34 gallons of water to produce a pound of potatoes, 222 gallons of water to produce a pound of pasta, and 518 gallons to produce a pound of chicken, it's easy to see that minimizing food waste by only buying what you need could be one of the most significant ways you can lower your water usage. (Boehrer, 2014)

- **Eat local, seasonal, quality food** - The water footprint of food is directly influenced by how it is farmed and how far it has to travel before being eaten. Processed foods have the highest water footprint. For example, that pound of potatoes that takes 34 gallons of water to produce becomes a pound of potato chips that takes 125 gallons! (Link, 2022)

- **Steam your veggies**—Instead of boiling vegetables, which uses lots of water, try steaming them. In addition to conserving water by steaming or pressure-cooking vegetables, more nutrients are retained.

- **Don't throw water out** - Instead of pouring leftover water in the bottom of cups and bottles down the drain, use it to water plants. I take a glass of water to bed with me each night, and whatever is left in the morning gets poured into the philodendron that lives in my room. After dinner, I do this with what's left in everyone's glass, watering the plants in the living room and kitchen. This small step ensures that my houseplants get the water they need without using any extra water.

- **Reuse boiling water**—When you cook pasta, eggs, or other boiled foods, drain them into a heat-safe bowl, allow the water to cool, and use it to water your garden. As long as you haven't added salt or other seasonings to the water, this is a great way to ensure you are getting the most use out of your water. As an added bonus, the starch in the water left from boiling pasta is excellent plant food! (Searle et al., 2023)

In the Garden

- **Check fixtures for leaks** - External leaks are much easier to miss than leaks inside your home because they usually leak onto grass or concrete. These leaks, however, can cause hundreds of gallons of waste each year. Make sure that faucets are turned completely off and not dripping. This is especially important when the water supply is kept on and the flow is regulated by a nozzle at the end of the hose.

- **Use compost in your garden** - Compost increases the soil's capability to absorb water. (Eartheasy, 2018) If you currently compost, make sure you're composting regularly. If you don't compost yet, this is another great reason to start! Mulching around landscaping and trees with compost or other organic material reduces the amount of water your yard needs.

- **Time your gardening** - watering plants when the sun is directly on them increases water waste because evaporation occurs more quickly, meaning plants need more water. Water in the early morning or late evening to maximize the water your plants receive. Additionally, try to deliver the water as close to the roots as possible. Water that falls on leaves will evaporate; the water that matters is what gets into the soil. A slow-drip irrigation system helps water reach the roots of the plants with less waste.

- **Plant in hydro-zones** - Grouping plants with similar water needs together can help you manage watering more efficiently. It allows you to focus on specific areas that need more water while ensuring that others don't get overwatered.

- **Plant drought-resistant plants** - Plants native to your area typically survive with less watering since they thrive in the natural conditions around you. Decorative landscaping with regional wildflowers and shrubs is becoming more common as awareness of water consumption increases. Succulents and other drought-resistant plants can also provide an attractive landscape that doesn't require intensive watering.

Around the Home

- **Sweep or blow off your walkways** - Instead of using a high-pressure water hose to clean your walkways, sweep them off or use a leaf blower to move debris. This requires a bit more work, but it's a clear water savings, and you get the extra bonus of exercise and fresh air.

- **Wash your car using a bucket**—Washing your car can use more than a hundred gallons of water (Eartheasy, 2018). You can minimize this by using a bucket of soapy water for washing and running the hose only to rinse the soap off your car.

- **Wash full loads of clothes** - Make sure you wash the largest loads your machine can comfortably handle. Even energy-conserving models will use more water to wash two small loads of clothes than one large one.

- **Cover swimming pools** - If you have a swimming pool of any size, covering it can minimize evaporation. Swimming pools can lose an inch of water per week to evaporation during the summer. (Eartheasy, 2018)

As you can see, there are many ways to increase the sustainability levels of our homes. Probably no one will implement all of these suggestions, but even using two or three can positively impact your water footprint. Let's take a look at the next pillar of a prepper mindset: resourcefulness.

Resourcefulness: Edible Water

Things get tricky when trying to make sure you're adequately hydrated while still working to conserve water. Luckily, there are a few techniques that can make it easy. Certain foods contain large amounts of water and electrolytes, which can help you stay hydrated during a water shortage.

What Are Electrolytes?

Electrolytes are minerals with an electric charge. They include sodium, calcium, magnesium, and phosphate. (Whelan, 2019) These minerals are in charge of many things in our bodies, from sending signals from our brains to our muscles to building new tissues. Our hearts keep beating because of signals from electrolytes, meaning they are literally keeping us alive. (Felman, 2023)

Unfortunately, water depleted of these minerals may not satisfy our thirst because our bodies still need the minerals they hoped to get from the water. Electrolyte deficiencies can lead to many of the same symptoms as dehydration (Whelan, 2019). Eating electrolyte-rich foods is a resourceful way to avoid this problem, whether or not there is a shortage of drinking water.

Foods Containing Electrolytes

Below are the five main electrolytes you need in your diet and a list of foods containing large amounts of that mineral. Some foods will be listed more than once because they are excellent sources of several different minerals. (Whelan, 2019)

Sodium	Chloride	Magnesium	Potassium	Calcium
pickles	canned tomatoes	fish	bananas	kale
canned tomatoes	soup	spinach	yogurt	yogurt
soup	olives	turkey	potatoes	sardines
chicken	lettuce	chicken	turkey	milk
nuts	tomatoes	tomatoes	chicken	almonds
canned beans	celery	avocados	oranges	olives
table salt		soybeans	tomatoes	chicken
		watermelon	avocados	oranges
		strawberries	broccoli	tomatoes
			almonds	broccoli
			watermelon	soybeans
			strawberries	tofu

Foods Containing Water

In addition to eating electrolytes, certain foods contain a large percentage of water and can help supplement your water intake. It's normal to get about twenty percent of your needed water from the foods you eat, and you can increase that percentage by intentionally eating foods with a high water content. Cucumbers, radishes, tomatoes, lettuce, watermelon, zucchini, summer squash, asparagus, and mushrooms are hydration powerhouse foods. Each of them is over ninety percent water by volume. (UCLAHealth, 2022) Some other high-water-volume foods are peaches, watercress, spinach, and strawberries. (Brennan, 2020) Skim milk and broth are great additions to foods that increase the water content and provide necessary electrolytes. (Elliot, 2017)

If you are facing a water shortage, remember that food and beverages containing caffeine or alcohol will encourage dehydration in your body. Anytime you are concerned about getting proper hydration, whether it's an emergency or just a hot summer day, alcohol and caffeine intake should be limited. (Palkovick, 2022)

When drinking water supplies run low, edible water can be the key to staying hydrated. However, emergencies occasionally have such devastating consequences that all the resourcefulness you possess cannot mitigate them. In that case, being prepared by knowing what dehydration looks and feels like and how to treat it makes all the difference.

Emergency Preparedness: Dehydration

Dehydration is potentially deadly. Your kidneys can begin to shut down in as little as 24 hours. (Boynton, 2022) Learning to identify the first stages of dehydration will equip you to respond early and avoid the severe problems associated with severe dehydration.

Signs of Dehydration

Mild Dehydration

This first stage is when the body has lost up to five percent of the water it should have internally. A person may stop sweating as the body tries to conserve water. (WebMD, 2017) If you are experiencing mild dehydration, you will notice you become thirsty, tired, and dry-mouthed. You may also begin to have a headache. (Boynton, 2022) If a baby goes longer than three hours with a dry diaper or a toddler has a dry diaper for eight hours, they are likely experiencing mild dehydration. (WebMD, 2017)

Moderate Dehydration

The second stage is when the body has lost between six and ten percent of its necessary water. (WebMD, 2017) At this point, a scale would show a weight change of up to ten pounds as water weight is lost. At this stage, a person may also begin to have a headache of increasing intensity. Kidney function slows down, allowing a build-up of metabolic waste products in the body. This build-up can lead to nausea and vomiting. Urine will decrease and be darker due to the kidneys conserving the body's water rather than releasing it. (Boynton, 2022) This can occur in a few hours, especially if you are in a hot climate or are exercising vigorously.

Severe Dehydration

The third stage of dehydration is when the body loses more than ten percent of its regular water volume. This usually takes over 24 hours but can occur more quickly in a hot climate or in children. When you are severely dehydrated, a continued decrease in kidney function causes toxins to build up in the bloodstream. This can begin to damage other organs, in extreme cases causing your heart to stop beating. (Boynton, 2022) As toxins build up, a dehydrated person may experience a complete lack of urination, dizziness, confusion, irritability, and an increased body temperature. You may notice that their eyes begin to look sunken, they may breathe unusually rapidly, have a racing heartbeat, or faint. (WebMD, 2017)

Causes of Dehydration

Besides insufficient water intake, there are other causes of dehydration that are important to guard against. The most common is gastrointestinal illness, with diarrhea and vomiting. This is one of the reasons cleaning potential pathogens out of your water supply is vital during a water shortage. Dehydration can occur more quickly in hot climates or during physical exertion because you lose excess water by sweating. Also, certain medicines have a diuretic effect and can cause you to become dehydrated. (WebMD, 2017) Children are more likely to become dehydrated, especially in hot weather, because their water reserves and mass-to-surface-area ratio are smaller. (Tuscon Medical Center, 2019)

Treating Dehydration

The treatment for dehydration depends on the stage you or a family member is dealing with. The more severely dehydrated someone is, the more likely they are to need emergency medical care. (Boynton, 2022)

Mild Dehydration

Treat mild dehydration by drinking water as soon as you realize you are thirsty. If a family member shows symptoms, encourage them to drink water or a rehydrating solution immediately. Avoid drinking soda, coffee, alcohol, or energy drinks. Eat foods with a high water and electrolyte content. For children, freezing electrolyte-containing flavored drinks into popsicles may encourage consumption. (Tuscon Medical Center, 2019)

Moderate Dehydration

If symptoms of dehydration have persisted or worsened to the point of moderate dehydration, it's important to allow your body to cool down. Stop any physical activity and rest indoors or in the shade. Drink slowly, since guzzling water at this stage can shock your body and cause vomiting. Ensure that what you drink contains electrolytes to replenish your water and mineral content. (Boynton, 2022) If dehydration is due to an ongoing gastrointestinal illness, seek medical care when you notice signs of moderate dehydration.

Severe Dehydration

Once at the level of severe dehydration, seek medical care as quickly as possible. Once toxins have started to build up in your body, IV fluids are the best way to rehydrate since they are less likely to cause a shock reaction. (Boynton, 2022)

If medical care is unavailable, sip a rehydration solution slowly to minimize your body's adverse reaction while rehydrating. Place cooling packs on your pulse points to help lower body temperature, which can help your body accept oral rehydration. (*Dehydration and Heat Stroke*, 2019)

Rehydration Solutions

There are several oral rehydration solutions, commonly known as ORS, that you can buy. These drinks contain electrolytes to help stabilize your body if you are experiencing mild to moderate dehydration. Sports drinks, like Gatorade, are designed to meet the rehydration needs of adults and are marketed to people who are exercising or working outdoors. Pedialyte is sold as an ORS to rehydrate children during sickness. Because of this, Pedialyte contains significantly less sugar and more electrolytes than Gatorade. (Lang, 2023) ORS also comes in drink packets that can be mixed into clean water. These can be easier to store or carry with you than premade drinks.

If you don't have access to store-bought ORS or prefer to make your own, there are various recipes you can use. Molasses or maple syrup can be substituted for sugar in any of these recipes to add more potassium. (Staubs & Michelet, 2014)

- **Water-based solution** - Add 1/4 to 1/2 teaspoon of salt and 2-4 teaspoons of sugar per liter of water. Adjust the sugar and salt amounts to taste. (UVA Nutrition, 2023) Keep the sugar level as low as possible because too much sugar can cause diarrhea, especially in children. (Staubs & Michelet, 2014)

- **Broth-based solution** - Use chicken broth made from a bouillon cube according to the package directions or canned chicken broth, not low-sodium. If you have homemade chicken broth, add a teaspoon of salt while heating it. This solution is less helpful in hot climates since it would likely be drunk warm instead of cold. (UVA Nutrition, 2023)

- **Tomato-based solution** - Canned tomato juice diluted with water is another simple recipe for homemade ORS. Use plain tomato juice, not V8 or another mix containing tomato juice, and mix two cups with one cup of water. (UVA Nutrition, 2023)

- **Juice-based solution** - Mix 3/4 cup of plain cranberry or orange juice with three cups of water and 1/2 teaspoon of salt. Coconut water can also be used as an

ORS. (Staubs & Michelet, 2014)

- **Cereal-based solution** - Cook 1/2 cup of baby rice cereal according to the directions on the box. After the cooked cereal has cooled, add two cups of water and 1/4 teaspoon of salt. Mix well and refrigerate before drinking. The mixture should be similar in texture to a milkshake. (UVA Nutrition, 2023)

Rationing in Critical Situations

In earlier chapters, we talked extensively about storing adequate water supplies. The Federal Emergency Management Agency, known as FEMA, recommends storing a two-week water supply for your household (FEMA, 2005). If this goal is unachievable, store as much as possible. Unfortunately, no matter how much water you store, it's impossible to know ahead of time whether you might need more.

Determining a sensible response when water supplies run low is difficult, but the first priority should be to decrease your need for water by reducing activity levels and staying cool. These steps can help stretch your water supply. However, it is not recommended that you ration your water supply. (FEMA, 2005) It is better to give yourself adequate water today and try to find more tomorrow. The consequences of dehydration include brain fog and confusion, and rationing water could put your body into mild to moderate dehydration for an extended amount of time. This inhibits your ability to successfully find water due to a less-than-optimal physical and mental state. (Zawalsky, 2016)

The Prepper Mindset in Action

When I lived along the Amazon River, there was a period when the water supply to our village was shut off. After a few days with no running water and no idea when the local government would get it going again, our village leader and his family decided that we would find our own solution. A little over a kilometer away was a village that still had running water, so we pooled our resources and bought over a thousand meters of water hosing. We connected the hoses and ran them from their village through the jungle to ours so that we had a clean water supply. It worked!

If we had waited for someone else to fix our water shortage, that story could have ended very differently. But we determined that we would figure it out. The key to thriving no matter the emergency is the determination to use your resourcefulness and knowledge to ensure your family's safety. Use the knowledge you have gained from the first 6 chapters of this book to take the following quiz and see how prepared you are for an emergency.

Water Survival Quiz

Answers can be found at the end of the chapter.

1. How many gallons of water should one person store for two weeks?
2. How much chlorine should be added to each gallon of water before storing it?
3. True or false: It's fine to drink straight from a rainwater barrel.
4. What is the largest bottle that should be used for SODIS?
5. True or false: An old charcoal filter is better than no filter.
6. Does it use more water to wash dishes by hand or run the dishwasher?
7. What types of plants are best for plant-based water extraction?
8. True or false: It's important to store water for drinking, washing, and flushing toilets.
9. Which two advanced purification systems can be used for desalination?
10. True or false: Freezing temperatures can damage your water supply.

Using the Prepper Mindset to Thrive

Choosing to lower your water consumption now is the beginning of thriving using a prepper mindset. This not only provides benefits in an emergency; it also offers immediate conservation benefits like reducing energy bills and saving money on food. Buying local, seasonal produce decreases your water footprint while increasing your exposure to high water and electrolyte-containing foods when needed most. The foods that provide the most edible water are in season during the summer, when hot temperatures increase the risk of dehydration. Learning to increase water intake with foods and create your own ORS empowers you to avoid and treat dehydration in an emergency. Water survival comes with challenges, but you can overcome them. Chapter 7 will discuss potential challenges and how to prepare for them.

Answers for the Water Survival Quiz

1. Fourteen gallons
2. 1/8 teaspoon
3. False
4. Two liters
5. False
6. By hand
7. Large-leafed, non-toxic plants in full sun
8. True
9. Reverse osmosis and distillation
10. True

7

Navigating Potential Challenges

You may encounter many defeats, but you must not be defeated. In fact, it may be necessary to encounter the defeats so you can know who you are, what you can rise from, and how you can still come out of it." — *Maya Angelou*

THIS CHAPTER WILL EXAMINE emergencies and challenges you may face as you create and implement a water survival plan. My goal is to equip you with solutions for these challenges so that if an emergency strikes, you are prepared to take appropriate action.

Responding to Contaminated Sources

Public water can get contaminated at its source in several different ways. This can be due to fertilizers from agriculture, waste from feedlots, chemicals from manufacturing, or overflows due to flooding. Wildlife and natural mineral content can also pose a pollution risk as water enters the municipal water system when decaying animals or animal feces and the erosion of rocks containing toxic minerals enter the water supply at an unanticipated point. Also, when a community's distribution system starts to age, contaminants can be released into the water by cracks in the pipes or other problems with the system. (CDC, 2022)

Pollution can occur before or after treatment. If pre-treatment contamination poses a health risk, it means that water has been polluted by a chemical or pathogen not treated by that system. Post-treatment contamination is always a health risk since the water is being delivered to consumers and is assumed to be safe for drinking at that point. Public water systems are required to test their water at regular intervals and inform consumers of any contamination. They do this by releasing water advisories. (CDC, 2022) These advisories may be issued to address ongoing contamination or after a problem has been fixed. Municipal water companies may also issue an emergency notice, usually a boil advisory.

Boil advisory notices usually do not give information about what contaminants have been found in the water supply. Typically, they state the water is contaminated and should be boiled before use for food preparation, teeth brushing, handwashing, or drinking. (CDC, n.d.) Here are a few necessary precautions if you receive a boil advisory notice.

- Dump all of the ice in your ice maker. If you want ice, make it with bottled or purified water.
- Wash fruits and vegetables with bottled or purified water to avoid contaminating them.
- Brush your teeth with bottled or purified water.
- Pets can also get sick from many pathogens found during a boil advisory. Dump your pet's water and refill their bowl with bottled or purified water.
- Make baby formula with bottled or purified water, and sterilize any bottles or nipples that were washed before the boil advisory notice.
- All dishes must be rinsed with purified water to avoid contamination. If you run your dishwasher, it must be at its hottest setting. Alternatively, wash your dishes normally, fill your sink with tap water, and add 1 teaspoon of household bleach per gallon. Allow dishes to sit in the water for at least a minute before air drying.

Do not use the dishes until they are completely dry. (CDC, n.d.)

Possible Boil Advisory Contaminants

Although they won't specify the contaminants that may be found in your water supply, boil advisory notices can be triggered by several pathogens. Any of the following may result in a boil advisory.

- Cryptosporidium
- Giardia
- Legionella
- Campylobacter
- Norovirus
- Hepatitis A virus
- Rotavirus
- Enterovirus
- Salmonella
- E. coli
- Shigella

Municipal water suppliers suggest boiling the water because this list includes pathogens resistant to chlorination. (See Chapter 4) Boiling or SODIS are effective ways to purify water containing chlorine-resistant pathogens.

In chapters 4 and 5, we discussed methods of purifying water in detail. Before a crisis, ensure you are prepared with supplies to implement your chosen purification method. This will enable you to protect yourself, your family, and your pets from waterborne diseases such as cholera, diarrhea, dysentery, hepatitis A, typhoid, and polio.

Purifying Water While Traveling

Whether you are camping, traveling to a country with a potentially unsafe water supply, or preparing a "bug-out bag" for an emergency, having clean water is vital during travel.

International Travel

When traveling to parts of the world where access to safe water is limited, even bottled water may be inadequately treated. (Backer & Hill, 2023) International travelers must be prepared to treat their own water if necessary. Experts recommend packing water treatment tablets containing either iodine or chlorine if you are traveling to locations where the safety of the local water is questionable. If possible, boiling water before using it for drinking, washing dishes, cooking, or brushing your teeth is recommended. If you travel often or plan to stay long-term in an area without a dependably safe water supply, a portable reverse osmosis or UV filter would be a sensible investment. (Backer & Hill, 2023) These are available online for approximately $100 at the time of writing.

Camping

Carrying bottled water while camping or hiking can get quite heavy. Water bottles equipped with filters can be taken along empty and filled with groundwater for a lighter backpack when you are camping or hiking for an extended period. There are also straws containing filters that are easy to carry and allow you to drink from any freshwater source. It's essential to research the product you buy thoroughly before buying it since many physical filters don't filter out all viruses or protozoa. (Backer & Hill, 2023) Lifestraw, Berkey, and Grayl are three brands that make filtered water bottles that filter out over 99% of pathogens.

When wilderness camping, plan your campsites with basic water safety in mind. The toilet area should be at least 200 feet (about 70 steps) downstream from your water source. Dig a hole and be sure to bury all waste. When you move on, fill the hole completely. (REI Staff, 2017)

Verifying that your toilet area is built far enough from your water source is vital. When I lived in the Amazon, a nearby village built a new latrine. Within weeks, the villagers began getting sick. After a few weeks of gastrointestinal problems, the village leaders realized that the new toilets had been built too close to the well and were leaching contaminants into the aquifer supplying the well. Shortly after they dug a new latrine further away from the well, everyone started feeling better.

Bug-Out Bag

When preparing a bug-out bag, you should follow many of the same basic principles as if packing for international travel or wilderness camping. Carrying enough bottled water to survive an emergency takes up a lot of space. Water is also heavy, so unless building your muscles is a pressing priority, it's wise to take portable water purification supplies instead. Including these in your bug-out bag assures you that you would be equipped to purify water in a life-threatening situation. The best plan is to carry a combination of water purification tablets and lightweight or portable filters. This will equip you to purify and drink water from any source you find.

Desalination and Uses for Salt Water

If you live near a coast, while you have plenty of access to water, you may be unable to use it due to the salt content. In this case, a reverse osmosis or distillation system would be an excellent investment. Either of these would allow you to harvest seawater and use it to make drinking water in an emergency. Small distillation systems are more affordable than similar reverse osmosis filters. However, they cost much more to operate because a distillation system uses about sixty times as much electricity as a reverse osmosis system. (Woodard, 2024)

To minimize the amount of water you need to desalinate in an emergency, it's important to know what salt water can be used for. Below are the ways that salt water can be utilized.

Before utilizing seawater for household uses, filter it through a tight mesh or homemade filter (instructions in Chapter 4) to remove physical impurities. The higher the seawater temperature, the more likely it is to contain harmful bacteria. In warm climates, seawater should be filtered through an activated carbon filter before you use it to brush your teeth or wash your hands or dishes. (Kusek, 2006)

- Laundry - Wash your clothes in a five-gallon bucket of seawater by filling the bucket approximately half full. Add concentrated liquid detergent. Scrub any stains with a bar of castile soap. Once the clothes are thoroughly wet, use a broom handle or something similar to agitate the clothes for five to ten minutes. Squeeze as much seawater as possible from the clothes. Put a few inches of desalinated water into another bucket and rinse one piece of clothing at a time by swishing it through the fresh water. (Sailing with Bloom, 2017)

- Toothbrushing - Brushing your teeth with seawater will produce less lather from your toothpaste, but the high sodium content is good for your gum health. (Brumbaugh, n.d.)

- Flushing toilets - Although continued saltwater use may harm your plumbing system, emergency use of seawater in your drains and toilets is acceptable. (*Saltwater FAQs*, 2024)

- Handwashing—Lye or castile soap is the best choice for washing in seawater because it doesn't leave a residue and creates suds, which many soaps won't do in saltwater.

- Bathing - If your skin is irritated by the salty residue after bathing, rinse your skin in a basin of desalinated water from your purification system. (Brumbaugh, n.d.) Castile soap can be used for bathing and washing hair.

- Washing dishes - You can wash dishes in warm salt water with a concentrated liquid soap. Then, rinse them with a few inches of desalinated water to rinse your dishes.

With a desalination system set up and knowledge of how to properly use seawater, you are prepared for a water emergency on an island or near the coast. Remember that seawater cannot be used for irrigation, drinking, or pets and livestock.

Greywater Recycling

Greywater is the term for water utilized for household uses, without coming into contact with feces or other waste products. Greywater, the waste from showers, handwashing, and laundry, makes up sixty percent of the wastewater from the average household. (World Wildlife Federation, n.d.) Capturing and recycling greywater is a great way to increase the amount of available water in an emergency.

Greywater can be reused for flushing toilets. Irrigation is another potential use of greywater, though it's crucial to consider what you're putting into it. Using natural and non-toxic cleaning products, laundry detergent, and soap is essential if you intend to water plants, especially food plants, with your greywater.

Designing a Greywater Recycling System

Simple Greywater Systems

Creating a greywater recycling system will enable you to reuse your water and conserve your water resources. The simplest method is to collect water by disconnecting the pipe under the kitchen or bathroom sink and placing a bucket under the pipe that comes from the sink.

You can flush toilets with this water by pouring it directly into the bowl. Do not pour it into the water tank, especially if you are using water from the kitchen sink, since kitchen greywater tends to have more pieces of organic matter and will likely clog the tank. (World Wildlife Federation, n.d.) Do not let greywater sit in the bucket for over 24 hours, or it may begin to smell. (Greywater Action, n.d.)

Water collected in buckets under your kitchen and bathroom sinks can also be used for irrigation. This takes a few more steps than flushing your toilet.

1. Carry the bucket of greywater outside.

2. Get a second bucket and securely place a piece of window screen over it.

3. Pour the greywater slowly through the screen to filter any fragments of food, hair, or other potential contaminants from the greywater. Whenever necessary,

stop and clean the screen of debris.

4. Pour the filtered greywater onto the soil next to the plants you are watering, not directly on the plant itself. Do not allow greywater to directly touch any edible part of the plants. (Greywater Action, n.d.)

Advanced Greywater System

It's only slightly more complicated to design a greywater system that connects your drain to your landscaping without your direct participation, so once it's set up, I'm guessing you'll agree it was worth the effort. Before starting this process, check state and local restrictions to ensure that attaching your kitchen sink to a greywater catchment system is permitted. In many places, recycling greywater from the kitchen sink is prohibited. (Allen, 2017)

Supplies

- 55-gallon barrel with a spigot near the bottom (the same as a rainwater barrel)
- Sand
- Crushed stone
- Gravel
- Weed mat
- Three-way valve or Y-valve
- Nylon hose and PVC pipe
- Garden hose
- Small nail

Instructions:

1. Divert the water from the drain of your washing machine, shower, or sink by installing a three-way valve or a Y-valve at the point where these drain pipes would typically drain to the sewer. Depending on your experience level with DIY projects, you may need a plumber's help.

2. Place the barrel where you intend to keep it. Raise it several inches off the ground so the garden hose can attach to the spigot without crimping.

3. Prepare your greywater barrel by placing six inches of crushed stone into the bottom.

4. Cover the crushed stone with a screen or mesh filter cut to fit the barrel.

5. Pour a ten-inch layer of sand on top of the screen.

6. Layer one inch of gravel on top of the sand.

7. Use nylon hose and PVC pipes to create a discharge hose that drains into the top of your barrel. The method for this will vary depending on your specific situation.

8. Push the nail completely through the garden hose to create two holes at each push point. Create these holes at three-inch intervals to create a soaker hose.

9. Attach the soaker hose to the spigot at the bottom of the greywater barrel and open the spigot.

10. Lay the soaker hose on your garden soil between the rows.

As water flows into the barrel and through the filters, it will flow into the hose and out through the holes you have made. This delivers water directly to the roots of the plants you are watering without splashing any onto the plants, wasting less and ensuring that edible plants are not splashed by greywater. (Caldwell, 2013)

Planning Effectively

One of nature's paradoxes is that beauty and danger often go hand-in-hand. I remember witnessing the most beautiful lightning storm I'd ever seen and thinking it was like watching a fantastic fireworks display. I was delighted until it caused a tree to crash down on the power lines near our house, knocking out our power source. While this wasn't a true emergency, it serves as a reminder that planning ahead is the best way to perform effectively in any situation, illustrated by my stubbed toe on the way to look for a flashlight.

When we lived on the Amazon, our village was across the river from the nearest town where we could buy supplies. Just north of our village was a broken bridge, which had been built only ten years earlier. At the time of building, it seemed like a great plan, enthusiastically embraced by the locals. The problem was that the plan failed to account for the river's power. After a few short months, the bridge's support beams crumbled from the ongoing force of seasonal flooding and the river's powerful current. So, within sight of an unusable broken bridge, we had to park our vehicles on the opposite side of the river and ferry supplies across in small boats. Driving across would have made life so much easier! However, inadequate planning and insufficient follow-through doomed the project to collapse, literally and figuratively. This is a stark reminder that enthusiasm and any old plan will not suffice, no matter how good our intentions are. We must also consider potential challenges and tackle them before it's too late.

This is especially true when we are planning for a water emergency. Recognizing potential challenges and finding solutions before you need to implement them is the best way to design a successful plan that will support your family through a crisis rather than crumble under pressure. In the next chapter, we will use all of the knowledge you've gained throughout this book to create a thirty-day emergency plan for you and your family. This way, you will have a pre-existing system and strategy in the unlikely event of a significant water emergency.

8

The 30-Day Emergency Plan

"To appropriately respond to an emergency requires a very clear mind, to cooly analyze what the observations are and how to fix it." — *Buzz Aldrin*

ARE YOU READY TO put together everything you've learned? This chapter will be about creating a personalized plan for your family so that you can be prepared if a water emergency strikes. By the end of this chapter, you'll know exactly how you can be ready within the next thirty days for any potential disaster that interrupts your water supply.

Week 1: Assessment and Strategy

Your task for the first week is to assess your water needs during a potential emergency and create a strategy to meet those needs. This should include discovering local agencies with information specific to your home area.

Step 1 - Make a Goal Statement

The first step is to decide what your goal is. Are you preparing to provide a thirty-day water supply for your family, or will you only gather enough water for two weeks? Would you prefer to stockpile a larger water supply or be able to source water during an emergency? Starting with a clear vision of how you want to handle a potential emergency will streamline your planning process.

Create a statement that sums your goal up in one sentence. Here are a few examples of potential goal statements:

- **Scenario 1**: I plan to store drinking water for two people and one dog for twenty-eight days, install a rainwater barrel in the backyard to cover our non-potable water needs, and install a small filtration system in case we need to drink rainwater.

- **Scenario 2:** Our goal is to drive a well on our property with a hand pump to meet our water needs in an emergency and install a water storage tank for our non-potable water needs.

- **Scenario 3**: We plan to purchase a reverse osmosis water filtration system, store two years' worth of replacement filters, and learn how to use seawater for non-potable water needs.

As you can see, this step isn't about figuring out how you will accomplish your goal; it's simply about setting a goal that allows you to begin preparing ahead for a potential emergency.

Step 2 - Review Your Drinking Water Needs

Now that you have established your goal, you can begin assessing what you need to meet it. Chapter 3 provided detailed instructions for calculating the amount of drinking water to store for people and animals. Use this resource now to figure out how much water you need to have on hand.

Having at least a few days of drinking water on hand is a good idea, even if your goal includes providing water from other sources during an emergency. That way, if you have to leave your home temporarily or encounter an unexpected complication with your water source, you have water for several days.

Using the three sample goal statements above, here are examples of how to figure out how much water to store:

- **Scenario 1**: Two people and a medium-large dog each need one gallon of drinking water per day for twenty-eight days.

3 people and dog x 28 days x 1 gallon = 84 gallons needed for storage

- **Scenario 2**: A family of four will drive a well on their property but wants to store a week's worth of water in small bottles in case they need to leave home in an emergency or have an issue with their well.

4 people x 7 days x 1 gallon = 28 gallons + 7 gallons extra because one member of the family is pregnant = 35 gallons needed for storage

- **Scenario 3**: A family of five that lives near the beach plans to use a reverse osmosis filter for emergency water purification and desalination. They also want to store two weeks of drinking water so that they have drinking water if their filter fails.

5 people x 14 days x 1 gallon = 70 gallons needed for storage

As you can see, this step is only focused on storing drinking water. For those with food gardens needing water in an emergency, storing water for irrigation is crucial.

Step 3 - Plan for Your Non-Potable Water Needs

Storing extra drinking water can fill non-potable water needs. This is only advisable in small families since storing the amount of water needed to meet the needs of a larger family may be costly and complicated.

A large water storage tank or rainwater barrel is a more cost-effective and sustainable way to provide non-potable water. A greywater recycling system would also supplement water for irrigation and sanitation.

Chapter 3 provides instructions for estimating the amount of water needed for laundry, cleaning, sanitation, and irrigation. Use this information to consolidate your non-potable water supply plan based on an assessment of your emergency needs.

Step 4 - Find Your Challenges

Local emergency organizations can provide information on what emergencies are considered most likely where you live and the average length of time that water supply would be interrupted in those situations. There's no time like the present to sign up for emergency alerts from local and national agencies to ensure that you have timely notice of a potential crisis.

Knowing what type of disasters are most likely in your area can help you decide if it makes more sense to evacuate in a disaster or prepare to stay home safely. It can also give you an idea of how long it will take for normalcy to return. For example, the National Hurricane Center gives the anticipated timeframe for power and water outages as several days to a week for a Category 1 storm, with expected outages lasting up to several months for a Category 5. (National Hurricane Center, 2012)

You will also need to determine whether you are preparing for potential man-made emergencies that are less likely but could occur anywhere. This includes events such as terrorist or EMP attacks and civil unrest that affect basic services. The timeframe for the return of regular service for these events is unknown, so if you want to be prepared for them, it is wise to have a long-term water supply plan in place.

Step 5 - List Your Supplies

Once you have a firm plan for how much water to store and your preferred method of sourcing and purifying water if necessary, make a list of supplies required to make it happen. This consists primarily of bottled water and collection supplies for your planned collection method. It also includes any permits, surveys, and blueprints necessary to complete your projects. Let's use Scenario 1 from the steps above as an example.

Scenario 1 Supply List:

- 12 5-gallon water jugs to be kept in the utility room (60 gallons)
- Shelves for utility room water
- 10 1-gallon water jugs to be kept in the bottom of the pantry

- 5 1-gallon water jugs to be split between bedroom closets
- 3 24-packs of 16.9 oz. bottles in the garage (9 gallons/3 days portable supply)
- Rain barrel with a spigot installed
- Blocks for rain barrel
- Tin snips
- Gutter connections
- Screening
- Water hose
- Carbon filtration pitcher with 2 extra filters

As you can see, this family's supply list includes not only their stock of water but also all associated supplies, such as shelves and the necessary equipment to install their rainwater barrel. It also details where the stored water will be kept and uses their available space to spread the water storage out in a way that makes sense in their home.

By the end of week one, you should develop a plan that details how you will accomplish the goal you made in the first step and a list of the supplies you need to source. If you will need more supplies than you can gather in a week, start with your stock of bottled water so that you are prepared for your immediate needs if an emergency strikes before you have your whole plan in place.

Week 2: Resource Gathering

To prepare for a crisis that may interrupt your water supply for a month or more, experts recommend preparing a plan that includes three separate water sources (Pipkin, 2020). This way, if one of them is no longer viable, you still have an adequate water supply. During week two, begin obtaining your water stock and supplies for backup sourcing and purification methods.

The three staples of week two are storing your stock of water, discovering and sourcing supplies for tapping into supplementary sources, and ensuring you have the necessary purification supplies to turn dirty water into drinkable water.

Storage

Chapter 3 covers storage solutions for bottled water in detail. Remember to store water in various areas of your home. This allows you to access it easily and minimizes the chances that your entire water supply will be destroyed even if part of your home is damaged. Using a mix of 5-gallon, 1-gallon, and smaller containers helps make your water supply practicable.

An outdoor tank might be necessary to store large quantities of water. You should make your outdoor tank using a food-safe barrel, inspect it regularly for cracks, and ensure that it doesn't freeze in winter. Drinking water stored in a large tank or drum should be sterilized and sealed until you are ready to use it.

Sources

Chapter 2 details the sources of water you can tap into to supplement your store of bottled water. Let's review them quickly.

- **Rainwater** - Collecting rainwater in a rainwater barrel, cistern, or bucket is usually the easiest way to supplement your non-potable water needs. Additionally, in an emergency, rainwater can be purified for drinking.

- **Groundwater** - Tapping into underground reserves of water by digging, driving, or drilling a well can give you access to otherwise unobtainable resources. Open groundwater, like a stream or lake, provides a reliable source of water that can be purified for drinking.

- **Dew and condensation** - Collecting dew, fog, and air conditioner runoff is another supplemental method that provides water for purification.

- **Plant extraction**—In an emergency, water can be extracted from plants in various ways. Whether you build a solar still, a collection pit, or use leaves that are still on trees, water can be obtained from plants with large green leaves.

Purification

Knowing how to purify water for drinking is an essential element of disaster preparedness. Chapters 4 and 5 explain different methods of purification and their effectiveness. Below is a summary of each of the main points.

- **Distillation**—Distilled water is some of the purest water available. Distillation is the process of boiling water until it evaporates, then condensing it in a cooling chamber. The impurities are unable to evaporate with the water, so the resulting water is sanitized. Salt is also unable to evaporate, making distillation an effective means of desalinating seawater for drinking. Distillation units use a substantial amount of electricity since they must both heat and cool the water.

- **Reverse osmosis** - The other method for desalination of seawater is reverse osmosis. This method uses pressure to force water through a semi-permeable barrier that filters out pathogens, toxic chemicals, metals, and salt. Reverse osmosis (RO) also filters minerals out of water, meaning that electrolytes are no longer present in water that has been filtered using RO and should be obtained from food or electrolyte drink packages.

- **Boiling—Boiling is an** effective and simple way to kill all known pathogens. It is the oldest method and still one of the most effective. Simply bring water to a full boil and keep it there for a full minute (three full minutes if you live above 5,000 miles above sea level). If you don't like the taste of boiled water, you can improve it by adding a pinch of salt as it's cooling.

- **Solar disinfection** - Solar disinfection, known as SODIS, uses UV rays to kill pathogens that can cause illness. This is an excellent technique for areas and situations where boiling water is impractical. Using clear plastic bottles that are 2-liter or smaller or Ziploc bags, place the water in full sunlight for six continuous hours or forty-eight hours if it's partly cloudy. Remember to take into account any climate considerations for your area.

- **Chlorination** - Chlorination is a simple way to kill or inactivate most pathogens that cause water-borne diseases. Use 1/8 teaspoon of unscented bleach per gallon of water to sanitize it and let it sit for at least fifteen minutes before drinking it. If you are using well water or water from any source that has not been chemically treated for long-term storage, it should be chlorinated before storing it.

- **Filtration** - Whether you purchase a filtration system or make your own, filtering water can purify toxins and pathogens out of your water supply. Each element of a filter works to rid your water of specific contaminants. With the correct elements, a filter can clean up to 99% of pathogens, metals, and toxic chemicals out of water. Filtration can also work in tandem with other methods, such as SODIS, to make them more effective.

Once you have sourced your water and other necessary supplies, it's time to begin putting your plan into action. Having your family on board and ready to help at this stage is a great benefit since the next step involves the physical labor of putting your system in place.

Week 3: Implementation and Testing

This is where the fun begins! This week, your focus will be setting up a water supply system and making sure it's operating correctly. Depending on which water sources you plan for, this step could take a few days to several weeks. Implementing the easiest part of your plan first is the best way to build momentum.

There are two main elements of this week's activity. The first is setting up and distributing your water storage. The second is setting up and testing your water supply system. If you want to prepare properly for a long-term water outage, you need two supply systems.

Water Storage Set-up

You need to decide if you plan to store drinking water that you buy bottled. The alternative is to bottle it yourself. Bottled water is readily available for purchase in 5-gallon and smaller containers. If your storage plans include 55-gallon barrels or larger tanks, those need to be set up where you want to store them before being filled.

- Buy and set up any shelves you need for your water jugs.
- Buy and place jugs of water where they will be stored. Write the date on each one. If you are bottling your own water, treat the bottles before sealing them for storage.
- Set up your large drinking water barrels or tanks. Make sure you are happy with your placement before you start to fill them. Calculate how much chlorine or iodine to add to your large tank carefully before treating your water.
- After filling your tank or barrel and treating the water, ensure that it is completely sealed to keep the water sanitary. Write the date it was filled and sealed in a conspicuous location on the tank.

Once you have your bottled drinking water stored, it's time to turn your attention to getting a system in place for sourcing water during an emergency.

Water System Components

Water systems will vary significantly, depending on what you are sourcing water for. Rainwater barrels used for irrigation are considerably more straightforward than driving a well that will provide water for an entire home, for example.

There are four main components of any water system that supplies drinking water. (Pradhikaran, 2012) The section below details the choices you have for each of the different components.

Sources

- **Atmospheric Water** - Rainwater or dew
- **Groundwater** - Wells that are commercially drilled, driven, or hand-dug
- **Surface** - Rivers, springs, lakes, ponds, streams, or oceans

- **Purchased** - Municipal water that is inadequately treated

Treatment Steps

- **Step 1** - Primary filtration to remove large particles
- **Step 2** - Sand, pebbles, cloth, or activated charcoal filtration to remove small particles and toxic chemicals
- **Step 3** - Chlorine, iodine, biological filtration, boiling, or SODIS to kill or inactivate disease-causing pathogens
- **Alternative Step** - Reverse osmosis or distillation takes the place of steps 2 and 3 and desalinates water simultaneously

Storage

- **Elevated Storage** - Raised storage is preferable for systems that use gravity to move water.
- **Ground Level Storage** - The most common type of storage, this is the easiest to set up, maintain, and run.
- **Underground Storage** - Water stored underground must be pumped out to be used

Distribution

- **Gravity** - Setting up a system from a high point of storage to a low point of distribution allows you to use gravity alone to move your water supply
- **Hand Pump** - A positive aspect of a pump that is manually operated is that it will work as usual in an energy crisis

- **Motorized Pump** - These are much easier to operate but need a backup source of power.

Water System Set-up

Let's examine the components of setting up a rainwater barrel for irrigation as a simple water system example.

- **Source:** Rainwater
- **Treatment:** Primary filtration only
- **Storage:** Elevated on blocks or a small platform
- **Distribution:** Gravity fed as water feeds in from the gutter at the top to the spigot at the bottom

Let's break down the components of a more extensive system that creates drinking water from untreated surface water.

- **Source:** Surface water piped from a local stream
- **Treatment:** Two steps -
    - Primary filter to remove large particles
    - Point-of-entry reverse osmosis system where the water enters the home
- **Storage:** A ground-level water tank is installed in the garage
- **Distribution:** A pump is installed to pull water from the stream to the water treatment system

And finally, here is an example of a system that is less extensive but equally effective.

- **Source:** Rainwater
- **Treatment:** Three steps -
- Primary filtration as it flows into the rainwater barrel

- Filtering through an activated carbon filter when it's brought inside
- Chlorination before storing
- **Storage:** First-stage storage elevated in the rainwater barrel, second-stage storage after filtration and chlorination in closed 1-gallon jugs in the home
- **Distribution:** Gravity-fed through the rainwater barrel

These are just three of the many possible ways to set up a water supply system for drinking water. Remember to check local and regional laws before beginning any work on your water supply system. Even simple systems such as rainwater barrels are subject to local laws and ordinances.

Testing Your Water System

Testing that your system works correctly is extremely important. Even a tiny oversight can derail your plan. When I installed my first rainwater barrel, I forgot to close the spigot at the bottom of the tank. A few days after the next heavy rain, I went out to water my plants from my barrel and discovered it was empty. The water had poured out the open spigot while it was raining. I could have had all of that water available for use if I had tested to make sure the faucet was in the closed position.

Once you have your collection system set up, it is important to run water through it to test how it works. You may discover minor problems that are easy to rectify or major fixes that need to be made. Either way, it is much better to find them at this stage than in an actual emergency.

Week 4: Refinement

Preparing your family for a water emergency is not a one-time event. While you now have the main components of your plan in place, you must be ready to respond to changes in your life, family, and home. Additionally, all of your preparations will need maintenance. This will enable them to continue working as well as when you first set them up.

This is the week you will prepare your system to continue to work correctly. The focus this week is on water quality testing, maintenance supplies, and replacement parts.

Water Testing

Detailed information about water testing is provided in Chapter 4. This week is a great time to test the quality of the water running through your water treatment system. That way, you'll be able to gauge whether or not your water is adequately treated for safe drinking.

Choose your test kit based on what you want to ensure is not present in your water supply. Starting with a low-cost test strip that gives a reading on over twenty potential contaminants at once is a good idea. From there, you can zero in on any problem areas you find or buy test kits to check for specific contaminants that were not included in the original test. Experts recommend that water is tested yearly, so it is a good idea to buy several tests ahead of time. (Howell Jr., 2023)

The most effective way to get an exact reading on the safety of the water you have treated is to send a sample to an EPA-certified lab. Some labs can test for over two hundred potential contaminants. (Howell Jr., 2023) However, if lab testing is out of your budget, home testing can be a valuable tool to ensure water safety.

Maintenance Supplies

Ensure that you have the tools necessary to maintain your water system. Depending on its complexity, you might need to purchase plumbing tools and learn to use them. Remember that in an emergency, plumbers and repair workers will likely be in short supply, so knowing how to maintain your system and keep it running smoothly by yourself may be the difference between a working system and a broken one when you truly need it.

Replacement Parts

Keeping filters on hand for your treatment system is an obvious step to be prepared for an emergency, but having other replacement parts is also wise. For example, suppose you are using an activated carbon filter pitcher as part of your treatment system. In that case, I'm sure you have already calculated how many filters you should have on hand. But if the pitcher that holds the filter breaks, the filters are useless. So, if it is an integral part of your plan, you should have an extra pitcher on hand as well.

Moving Forward

Conduct a yearly review of your water emergency plan. Use and replace your stored water. Recalculate your stored water needs, which will change with family size and new pets. Change any filters that are nearing the end of their life. Perform recommended maintenance on your water system, checking it for leaks and replacing any seals, valves, or fittings that are showing signs of age.

This Is Just a Drill

An enlightening exercise to do with your family is a "water scarcity drill." Once you have your water emergency plan in place, plan a weekend or a few days when your family will practice having no running water.

Step 1: Turn your water off at its entry point.

Step 2: Walk through each daily scenario. Don't take the easy way out by going the whole weekend without a bath. Instead, figure out how to bathe with your stored or sourced water. Flush the toilets using the greywater that you just bathed in. You get the idea, right? Live as though the water will be out indefinitely. If you have kids, make this fun instead of acting like it's an inconvenience. Kids love an adventure!

Step 3: Take notes on what parts of your plan you would like to improve, then work on making them easier and better.

Step 4: A few weeks or months later, try it again with the improvements you've made. The more often you practice this drill, the more comfortable your whole family will be in a real emergency.

Now What...

Congratulations! You have now got an emergency plan for water scarcity. You have received the knowledge to conserve, harvest, purify, and store the most precious resource on Earth. Hopefully, you have been putting what you've learned into action as you've been reading. But if you still need to start, there's no time like the present to begin putting together what you've learned and translating it into a well-formed strategy that you can start implementing in the next thirty days.

As you work on incorporating these suggestions into your daily life, you will likely experience hiccups along the way. Don't worry; these are a natural part of the learning process! Keep practicing, and it will get easier and easier until water conservation is second nature and your emergency water plan is fully in place. This will give you peace of mind, knowing that in a water scarcity emergency, your family will not just survive. You will thrive!

Conclusion

"Thirst was made for water; inquiry for truth." — C.S. Lewis

We have come to the end of our journey. Do you feel prepared to navigate yourself and your family through a water scarcity emergency? The research I've shared here has equipped me with knowledge and the ability to understand water scarcity and create a plan to harvest, store, and purify my own supply of emergency water confidently. I hope you, too, are feeling a new surge of information-based confidence to prepare for a water crisis.

Looking at What You've Gained

Understanding the reasons behind the growing worldwide water shortage is the first step to becoming part of the solution. With the knowledge and ideas for water conservation you've learned by reading this book, you can take immediate action to reduce your water footprint.

A sustainability mindset, emergency preparedness, and resourcefulness will enable you to look at the challenges of preparing for a potential water emergency without worry. You won't be drawn into the fear cycle on the news or social media because you now have the practical knowledge to conquer the challenges and thrive no matter what happens.

Also, with your newly developed plan implemented, you can rest assured that a water crisis will not derail your family's health and well-being. With the tools you've gained, you are ready to meet and defeat whatever water emergency arises.

Looking at What's Next

The beautiful thing about learning new skills in one area is that your confidence to take on other challenges increases with each skill learned. Now that you are ready for a water scarcity event, what would you like to learn next? You may want to know how to forage for food or use natural medicine to heal yourself and your family. Bringing a prepper's mindset to each topic allows you to dive right in with a belief that you can make the world a better place and secure your family's future.

Also, if you have enjoyed this book, I would love it if you'd leave a positive review. This is a great way to help spread the message! More people need to hear that we don't need to be afraid or confused. Instead, we need to learn practical information about how to be ready, gaining tools and insights to successfully prepare for a water emergency.

Looking Way Back

In the ancient world, digging a well was a massive project, taking weeks or even months to finish. Jacob, a patriarch of the Jewish, Christian, and Muslim faiths, dug several wells, one of which still exists today in the city of Samaria in Israel. Jacob's Fountain, or the Well of Sychar, is best known for a significant encounter that happened there over a thousand years after it was dug. One of Jacob's descendents, Jesus, and a local woman who had come to fill her waterpot, had a lengthy discussion about water, life, faith, and choices.

Perhaps you were mystified or confused by some aspect of water survival before reading this book. Well, Jesus said something about water that made the woman scratch her head, puzzled by what He could be talking about:

"Everyone who drinks of this water will be thirsty again, but whoever drinks of the water that I will give him will never be thirsty again. The water that I will give him will become in him a spring of water welling up to eternal life."

It's easy to understand why the Samaritan woman was confused, right? How is it possible to never be thirsty again?

Everyone knows that water is a daily need for human beings. Without its life-sustaining properties inside of us, we cannot survive. We must plan and prepare beforehand to ensure that we have water in a crisis. Free water won't magically appear.

But Jesus wasn't talking about physical thirst or liquid water. He was talking about a deep internal thirst that each of us is born with. Do you ever wonder if there is something deeper in life?

Have you ever felt you were meant for more, searching for significance? This could be a subtle dissatisfaction that you push to the side or an intense desire for something you can't put your finger on. Perhaps it is a feeling that you're missing an important puzzle piece that could help things fall into place.

That's the thirst Jesus was talking about. It is a thirst for Jesus and His love and truth.

Jesus used the idea of water to illustrate that our deepest spiritual needs can be met through Him. Jesus offers us a relationship with God that brings eternal life to absolutely anyone who believes in Him. Just as physical water sustains our bodies, Jesus sustains our spiritual lives. While there are many temporary and partial fixes, Jesus is the only one who can satisfy our deep spiritual thirst completely and forever.

My earnest desire for you is that you would possess an adequate supply of physical water and also this living water that Jesus provides.

Feel free to contact me if you have any questions about Jesus or eternal life or if you'd like me to pray for you. My email is foraging@hadassahlewis.com.

And if you want to learn more about who Jesus is on your own, the book of John is a great place to start. https://biblia.com/bible/esv/john This section of the Bible is only about 16,000 words, half the length of the book you just read.

For those of you who enjoy connection and spiritual conversations, I'd love to hear anything you want to share about your opinions on faith, emergency preparedness, or water survival. Whether your perspective is similar to mine or 100 miles apart doesn't matter. I'm truly grateful that you took the time to read this book, including a couple personal stories I shared, and it would be an honor to learn about your journey and beliefs as well.

Call to Action

So now, after 8 chapters of diving into the technicalities of water survival together, you're ready to roll! With the knowledge you've gained, you are prepared to thrive in a water scarcity crisis. I applaud your initiative to gather all the necessary information to create a comprehensive water plan.

Now it's time to just do it! Gather your family, talk through sustainable choices, and create your personalized emergency plan to source, purify, and store the water you will need. Then, implement that plan over the next 30 days and beyond! I guarantee you won't be sorry.

One Last Thing

My purpose for writing this book is to help everyone develop their own water survival plan, and the only way for me to accomplish this is by reaching...well...everyone.

This is where you come in. Most people do, in fact, judge a book by its cover (and its reviews). So here's my ask on behalf of a struggling prepper you've never met:

Please help them out by leaving this book a review.

Your gift costs no money, only about 100 seconds of your time, but can change a fellow prepper's life forever. Your review could help...

- one more family find safe water in an emergency.
- one more community prepare for a disaster.
- one more person feel confident about water safety.
- one more child stay healthy during tough times.
- one more dream of safety come true.

By taking just a minute or two to leave an honest review on Amazon, you'll be pointing other individuals and families to this book. In doing so, you'll pass along information that equips them to collect, purify, and store the water they need for daily use and crisis situations.

Your insights and experiences can make a big difference in someone else's emergency survival plan. This knowledge has the potential to save lives, literally, from the deadly dangers of dehydration or a contaminated water supply.

Simply scan the QR code below to or click on the link to leave your review and make a difference:

[https://www.amazon.com/review/review-your-purchases/?asin=BOOKASIN]

Thank you for your support and for being a part of this sustainable, life-seeking community!

References

Adesina, T., & Beadle, R. (2020, March 19). *8 myths about clean water around the world*. ReliefWeb. https://reliefweb.int/report/world/8-myths-about-clean-water-around-world

Alberta Health Services. (2021, February 18). *Using disinfectants to make drinking water safe when you can't boil it*. My Health Alberta. https://myhealth.alberta.ca/Alberta/Pages/Using-disinfectants-to-make-drinking-water-safe-when-you-cant-boil-it.aspx

Alderson, C. (2021, February 24). *Wooden rain barrels vs plastic rain barrels pros and cons*. GardensAll. https://www.gardensall.com/wooden-rain-barrels/

Allen, L. (2017, March 1). *Kitchen greywater: Water conservation straight from the sink*. Hachette Book Group. https://www.hachettebookgroup.com/storey/kitchen-greywater-water-conservation-straight-from-the-sink/

Atlas Scientific. (2022, June 17). *The importance of groundwater*. Atlas Scientific Environmental Robotics. https://atlas-scientific.com/blog/the-importance-of-groundwater/

Auriga Research. (2021, March 29). *Types of water testing & how to test water*. Arbro Pharmaceuticals. https://aurigaresearch.com/water-testing/types-of-water-testing/

Backer, H., & Hill, V. (2023, May 1). *Water disinfection*. Centers for Disease Control and Prevention. https://wwwnc.cdc.gov/travel/yellowbook/2024/preparing/water-disinfection

Biome. (2018, March 18). *How to filter & remove microplastics from tap water*. Biome. https://www.biomestores.com/blogs/eco-home/remove-microplastics-tap-water

Blair, C. (2023, March 15). *How to keep water storage from freezing outside: Useful tips*. EZ-Prepping. https://ezprepping.com/how-to-keep-water-storage-from-freezing-outside/

BluGlacier. (2020, July 15). *Why is sustainability important?* BluGlacier. https://bluglacier.com/why-is-sustainability-important/

Boynton, E. (2022, August 17). *How dehydration affects the body*. UW Medicine. https://rightasrain.uwmedicine.org/body/exercise/dehydration

Brumbaugh, C. (2022). *How is hygiene possible with ocean water only?* The Great Outdoors Stack Exchange. https://outdoors.stackexchange.com/questions/4114/how-is-hygiene-possible-with-ocean-water-only

Caldwell, N. (2013, July 31). *DIY simple greywater system.* Better Farm. https://www.betterfarm.org/blog/2013/07/diy-simple-graywater-system.html

Campbell, B. (2022, March 1). *What is water distillation?* Waste Water Digest Magazine. https://www.wwdmag.com/what-is-articles/article/10940138/what-is-water-distillation

CBS News. (2011, June 21). *Guide to having safe drinking water after a storm.* CBS Miami. https://www.cbsnews.com/miami/news/keeping-your-water-supply-from-being-tapped-out/

CDC. (2008). *Household water treatment options in developing countries: CDC ceramic filtration.* Centers for Disease Control and Prevention. https://www.cdc.gov/safewater/publications_pages/options-ceramic.pdf

CDC. (2019). *Rainwater collection.* Centers for Disease Control and Prevention. https://www.cdc.gov/healthywater/drinking/private/rainwater-collection.html

CDC. (2020a, February 24). *Making water safe in an emergency.* Centers for Disease Control and Prevention. https://www.cdc.gov/healthywater/emergency/making-water-safe.html

CDC. (2020b, August 4). *Technical information on home water treatment technologies.* Centers for Disease Control and Prevention. https://www.cdc.gov/healthywater/drinking/home-water-treatment/household_water_treatment.html

CDC. (2022, May 26). *Water contamination and diseases.* Centers for Disease Control and Prevention. https://www.cdc.gov/healthywater/drinking/contamination.html

CDC. (n.d.). *Drinking water advisory communication toolbox.* Centers for Disease Control and Prevention. https://www.cdc.gov/healthywater/emergency/dwa-comm-toolbox/before/tools/FAQ-Boil-Water-Advisories-508.pdf

CFSS. (2024, February 20). *How does a commercial water filtration system work?* Central Florida Store Services. https://cflss.com/how-does-commercial-water-filtration-system-work-2/

Chapple, K. (2021, May 29). *Pros and cons of owning a rain barrel on the homestead.* Homestead.org. https://www.homestead.org/ecology/pros-and-cons-of-owning-a-rain-barrel-on-the-homestead/

Cho, R. (2011, July 1). *Wasting food = wasting water.* State of the Planet; Columbia Climate School. https://news.climate.columbia.edu/2011/07/01/wasting-food-wasting-water/

Chu, J. (2021, March 25). *MIT engineers make filters from tree branches to purify drinking water.* Massachusetts Institute of Technology. https://news.mit.edu/2021/filters-sapwood-purify-water-0325

Crystal Clear Water. (2022, December 5). *The surprising cons and pros of a reverse osmosis water filtration system.* CrystalClear-Systems. https://crystalclear-systems.com/pros-cons-reverse-osmosis-water/

CTCN. (2016, November 8). *Solar water disinfection.* Climate Technology Centre & Network. https://www.ctc-n.org/technologies/solar-water-disinfection

Cunningham, R. (2022, June 18). *How to store water long term in 4 easy steps*. Survival World. https://www.survivalworld.com/water/how-to-store-water-long-term/

Cunningham, R. (2023, May 2). *How to collect dew water: Efficient techniques for natural resource utilization*. Survival World. https://www.survivalworld.com/water/gathering-dew/

Davies, C. M., Roser, D. J., Feitz, A. J., & Ashbolt, N. (2009, February 1). *Solar radiation disinfection of drinking water at temperate latitudes: Inactivation rates for an optimized reactor configuration*. United States Environmental Protection Agency. https://cfpub.epa.gov/si/si_public_record_report.cfm?Lab=NERL&dirEntryId=205243

Dehydration and heat stroke. (2019). Johns Hopkins Medicine. https://www.hopkinsmedicine.org/health/conditions-and-diseases/dehydration-and-heat-stroke

dePeyster, E. (2014). *How much water does my food garden need?* Master Gardeners of Sonoma County. https://ucanr.edu/sites/scmg/files/185639.pdf

Eartheasy. (2018). *45+ ways to conserve water in the home and yard*. Eartheasy Guides & Articles. https://learn.eartheasy.com/guides/45-ways-to-conserve-water-in-the-home-and-yard/

Eggers, M. (2019, August 14). Infectious disease management and control with povidone iodine. *Infectious Diseases and Therapy*, *8*(4). https://doi.org/10.1007/s40121-019-00260-x

Eileen. (2022, July 22). *4 creative ways to collect rainwater without a gutter*. Just Measuring Up. https://www.justmeasuringup.com/collect-rainwater/

Elliot, B. (2017, August 9). *19 water-rich foods that help you stay hydrated*. Healthline. https://www.healthline.com/nutrition/19-hydrating-foods#TOC_TITLE_HDR_8

Emergency Essentials. (2013, October 1). *5 myths about water storage*. LDS Living. https://www.ldsliving.com/5-myths-about-water-storage/s/73851

Emergency water storage. (2020, February 28). Regional Water Providers Consortium. https://www.regionalh2o.org/emergency-preparedness/emergency-water-storage

EMIS. (2010, February). *Sand filtration*. Emis. https://emis.vito.be/en/bat/tools-overview/sheets/sand-filtration

EPA. (2015, November 18). *Emergency disinfection of drinking water*. Environmental Protection Agency. https://www.epa.gov/ground-water-and-drinking-water/emergency-disinfection-drinking-water

EPA. (2015). *Driven wells*. Environmental Protection Agency. https://www.epa.gov/sites/default/files/2015-11/documents/drivenwell_1.pdf

EPA. (2016, October 14). *Showerheads*. Environmental Protection Agency. https://www.epa.gov/watersense/showerheads

EPA. (2018, December 18). *Learn about private water wells*. Environmental Protection Agency. . https://www.epa.gov/privatewells/learn-about-private-water-wells

EPA. (2019, February 19). *What are volatile organic compounds (VOCs)?* Environmental Protection Agency.

https://www.epa.gov/indoor-air-quality-iaq/what-are-volatile-organic-compounds-vocs

Felman, A. (2023, January 6). *Electrolytes: Uses, imbalance, and supplementation.* Medical News Today. https://www.medicalnewstoday.com/articles/153188

FEMA. (2005, July 15). Food and water in an emergency. *Water Encyclopedia, 477.* https://doi.org/10.1002/047147844x.wr2172

FEMA. (2017, October 8). *Fact Sheet: How to Make Your Water Safe to Drink.* Federal Emergency Management Agency. https://www.fema.gov/press-release/20210318/fact-sheet-how-make-your-water-safe-drink

Firth, M. (2017, May 23). *Rain harvesting plumbing: Wet vs dry systems.* Enduraplas. https://blog.enduraplas.com/water-storage-rain-harvesting/rain-harvesting-plumbing-wet-vs-dry-systems

Franks, G. (n.d.). *Carbon filtration: What it does, what it doesn't.* Pure Water Products, LLC. https://www.purewaterproducts.com/articles/carbon

Garcia, M. (2024, January 24). *How to store water long term.* WikiHow. https://www.wikihow.com/Store-Water-Long-Term

García-Gil, Á., García-Muñoz, R. A., McGuigan, K. G., & Marugán, J. (2021, June 5). Solar water disinfection to produce safe drinking water: A review of parameters, enhancements, and modeling approaches to make SODIS faster and safer. *Molecules, 26*(11), 3431. https://doi.org/10.3390/molecules26113431

Gonzalez-Cebrian, P. (2019, September 6). *What are today's water challenges?* Smart Water Magazine. https://smartwatermagazine.com/blogs/pablo-gonzalez-cebrian/what-are-todays-water-challenges

Gorin, A. (2022, March 1). *How much water is too much?* Food Network. https://www.foodnetwork.com/healthy/articles/is-there-such-thing-as-drinking-too-much-water

Greywater Action. (n.d.). *About greywater reuse.* Greywater Action. https://greywateraction.org/greywater-reuse/

Hague Water Supply. (2019, August 26). *Pros and cons of UV filtration.* Hague Quality Water of MD. https://haguewaterofmd.com/pros-and-cons-of-uv-filtration/

Hancock, N. (2017, January 23). *Safe drinking water fact sheet.* Safe Drinking Water Foundation. https://www.safewater.org/fact-sheets-1/2017/1/23/what-is-chlorination

Health Vermont. (2024, February). *After a flood: Private drinking water guidance.* Vermont Department of Health. https://www.healthvermont.gov/environment/drinking-water/after-flood-private-drinking-water-guidance

Howell Jr., M. (2023, July 3). *How to test your water quality at home in 5 steps.* Consumer Notice, LLC. https://www.consumernotice.org/environmental/water-contamination/testing-water/

Innovative Water Solutions. (2023). *Rainwater harvesting 101.* Innovative Water Solutions. https://www.watercache.com/education/rainwater-harvesting-101

Jones, G. T. (n.d.). *Desert survival: How to build a solar still.* DesertUSA. https://www.desertusa.com/desert-people/water-solar-still.html

Juvakoski, A., Singhal, G., Manzano, M. A., Moriñigo, M. Á., Vahala, R., & Levchuk, I. (2022, June 25). Solar disinfection – An appropriate water treatment method to inactivate fecal bacteria in cold climates. *Science of the Total Environment*, *827*, 154086. https://doi.org/10.1016/j.scitotenv.2022.154086

Karanja, N. (2023, March 1). *How much water should you drink per day based on your age and gender?* LinkedIn. https://www.linkedin.com/pulse/how-much-water-should-you-drink-per-day-based-your-age-njeri-karanja/

Kershner, K. (2008, May 8). *How reverse osmosis works.* HowStuffWorks. https://science.howstuffworks.com/reverse-osmosis.htm

Kiger, P. (2013, April 15). *10 innovations in water purification.* HowStuffWorks. https://science.howstuffworks.com/environmental/green-tech/sustainable/10-innovations-water-purification.htm

Klobucista, C., & Robinson, K. (2023, April 3). *Water stress: A global problem that's getting worse.* Council on Foreign Relations. https://www.cfr.org/backgrounder/water-stress-global-problem-thats-getting-worse

Kusek, K. (2006, September 13). *The oceans are filled with microbes. Some are nasty.* Oceanus. https://www.whoi.edu/oceanus/feature/the-oceans-are-filled-with-microbes-some-are-nasty/.

Lancaster, B. (2024). *Condensate & dew harvesting.* Rainwater Harvesting for Drylands and Beyond. https://www.harvestingrainwater.com/water-harvesting/harvests-of-different-waters/condensate-harvesting/

Lang, A. (2023, July 28). *Pedialyte vs. Gatorade: What's the difference?* Healthline. https://www.healthline.com/nutrition/pedialyte-vs-gatorade#main-differences

Lawson, J. A., & Mistry, P. (2017, July 13). *How to test water quality? Chemical tests for limited budgets.* World Bank Blogs. https://blogs.worldbank.org/en/water/how-test-water-quality-chemical-tests-limited-budgets

Lindholm, H. (2019, April 2). *The foundation of life – Rainwater harvesting throughout history.* Bushman Tanks. https://www.bushmantanks.com.au/blog/the-foundation-of-life-rainwater-harvesting-throughout-history/

Link, K. (2022, March 7). *What you eat has a water footprint.* FoodPrint. https://foodprint.org/blog/what-you-eat-has-a-water-footprint/

Liu, C. (2022, April 8). *95% of U.S. tap water contains microplastics.* Hydroviv. https://www.hydroviv.com/blogs/water-smarts/95-us-tap-water-contains-microplastics?nbt=nb%3Aadwords%3Ax%3A15393145424%3A%3A&nb_adtype=&nb_kwd
=&nb_ti=&nb_mi=&nb_pc=&nb_pi=&nb_ppi=&nb_placement=&nb_si=

Living Whole. (2024, March 23). *Do water distillers use a lot of electricity?* Living Whole. https://www.livingwhole.com.au/do-water-distillers-use-a-lot-of-electricity/

Lone Star Drills. (2018, May 25). *Clues for finding groundwater.* Lone Star Drills. https://www.lonestardrills.com/clues-for-finding-groundwater/#:~:text=Existing%20hand%2Ddug%20wells%20will

Lopez, M. J., & Hall, C. A. (2023, March 13). *Physiology, osmosis.* National Library of Medicine. https://www.ncbi.nlm.nih.gov/books/NBK557609/

Luzi, S., Tobler, M., Suter, F., & Meierhofer, R. (2016). *Sandec: Sanitation, water, and solid waste for development SODIS manual guidance on solar water disinfection.* https://www.sodis.ch/methode/anwendung/ausbildungsmaterial/dokumente_material/sodismanual_2016.pdf

MacDougall, L. (2011, April). *Drive your own freshwater well.* Backwoods Home Magazine. https://www.backwoodshome.com/drive-your-own-freshwater-well/

MacGill, M. (2021, November 18). *How much water should I drink each day?* Medical News Today. https://www.medicalnewstoday.com/articles/306638#intake-by-age

Macrotrends. (n.d.). *Sao Paulo, Brazil metro area population 1950-2022.* Macrotrends. https://www.macrotrends.net/cities/20287/sao-paulo/population

MacWelch, T. (2015, July 10). *5 things your pets need to survive an emergency.* Outdoor Life. https://www.outdoorlife.com/blogs/survivalist/5-things-your-pets-need-survive-emergency/

Malegowski, V. (2017). How to collect clean, drinkable water from plants [YouTube Video]. In *YouTube.* https://www.youtube.com/watch?v=yQLXqyKlxKY

Maxwell-Gaines, C. (2018, May 5). *What are the benefits and advantages of rainwater*? Innovative Water Solutions, LLC. https://www.watercache.com/faqs/rainwater-harvesting-benefits

McBride, J., & Berman, N. (2023, March 31). *How U.S. water infrastructure works.* Council on Foreign Relations. https://www.cfr.org/backgrounder/how-us-water-infrastructure-works

Meehan, M., Stokka, G., & Mostrom, M. (2015, July 19). *Livestock water quality.* NDSU Agriculture and Extension. https://www.ndsu.edu/agriculture/extension/publications/livestock-water-quality

Meierhofer, R. (2002). *Solar water disinfection: A guide for the application of SODIS.* https://www.researchgate.net/publication/285888824_Solar_Water_Disinfection_-_A_guide_for_the_application_of_SODIS

Merriam-Webster. (2024, January 14). *Definition of RESOURCEFULNESS.* Merriam-Webster. https://www.merriam-webster.com/dictionary/resourcefulness

Michaud, K. (n.d.). *Homemade water filter science project.* LoveToKnow. https://www.lovetoknow.com/parenting/kids/homemade-water-filter-science-project

Microfarm Garden. (2019, January 10). *The history of rainwater harvesting.* Microfarmgardens.com. http://microfarmgardens.com/blog/2019/1/10/the-history-of-rainwater-harvesting.html

Middendorf II, J. W. (2023, January 2). *Blackout shows how fast the US would fall in EMP attack.* Energy Central. https://energycentral.com/news/blackout-shows-how-fast-us-would-fall-emp-attack

Migala, J. (2015, February 16). *The health benefits of water.* EverydayHealth. https://www.everydayhealth.com/water-health/water-body-health.aspx

Mother Earth News. (2022, May 3). *Tips for using a manual water pump.* Mother Earth News.

https://www.motherearthnews.com/homesteading-and-livestock/self-reliance/manual-water-pump-zm0z22jjzols/

Mullen, K. (2019). *Groundwater.* NGWA. https://www.ngwa.org/what-is-groundwater/About-groundwater/information-on-earths-water

Neveln, V. (2022, March 3). *Make your own rain barrel from a garbage can in 5 easy steps.* Better Homes & Gardens. https://www.bhg.com/gardening/design/projects/how-to-make-rain-barrel/

NHC. (2012). *Saffir-Simpson hurricane wind scale.* National Hurricane Center. https://www.nhc.noaa.gov/aboutsshws.php

NSF. *(2009). Publication No. 7: Distillation treatment systems for private drinking water system.* Connecticut Department of Environmental Health. https://portal.ct.gov/-/media/Departments-and-Agencies/DPH/dph/environmental_health/pdf/07DistillationTreatmentSystforPDWSpdf.pdf

Oakville Water Services. (2019, September 29). *What is the best location for my water storage tank?* Oakville Pump Service. https://oakvillepump.com/blog/2019/9/29/what-is-the-best-location-for-my-water-storage-tank

Palkovick, L. (2022, July 27). *How to recover from alcohol- or coffee-induced dehydration.* Drip Hydration. https://driphydration.com/blog/alcohol-and-coffee-induced-dehydration/

Patel, H. H. (2018, May 14). *What is activated charcoal?* News-Medical. https://www.news-medical.net/health/What-is-Activated-Charcoal.aspx

Peterson, C. (2023, November 29). *How long can you survive without food?* Outdoor Life. https://www.outdoorlife.com/survival/how-long-can-you-survive-without-food/

Pickett, E. (2022, August 12). *Free water for plants if you have an air conditioner.* Youtube. https://www.youtube.com/watch?v=EXsfgbi7OO8

Pipkin, C. (2020, September 17). *A to Z emergency prep guide: How to survive for 30 days & beyond.* Be Prepared - Emergency Essentials. https://www.beprepared.com/blogs/articles/a-to-z-guide-1-month-beyond

Pradhikaran, J. M. (2012). *Module 1 basics of water supply system -training module for local water and sanitation management.* CEPT University. https://www.pas.org.in/Portal/document/ResourcesFiles/pdfs/Module_1%20Basics%20of%20water%20supply%20system.pdf

REI. (2017, October 17). *How to go to the bathroom in the woods.* Recreational Equipment, Inc. https://www.rei.com/learn/expert-advice/hygiene-sanitation.html

Rosa e Silva, G. O., Loureiro, H. O., Soares, L. G., de Andrade, L. H., & Santos, R. G. L. (2021, November 21). Evaluation of an alternative household water treatment system based on slow filtration and solar disinfection. *Journal of Water and Health, 20*(1), 157–166. https://doi.org/10.2166/wh.2021.211

Rosario, O., Machado, P., Kennedy, J., & Ana Del Yacuma, S. (2022, July). *Banana peel for water purification.* https://watertank.siwi.org/wp-content/uploads/2022/07/uni2-x-el-agua-team-bolivia.pdf

Royer, K. (n.d.). *UV water purification: Ultra-violet, ultra-effective.* Aquasana. https://www.aquasana.com/info/uv-water-purification-ultra-violet-ultra-effective-pd.html?discountcode=LIVEHEALTHY&gclid=CjwKCAjwi_exBhA8EiwA_kU1Mge9srvGDbpfxB8YzA9RXnutvZyRpNPLKZEIHxeERI_oQKBxamJcPBoCN08QAvD_BwE&gad_source=1

RTS. (n.d.). *Food waste in America in 2020: Statistics & facts.* Recycle Track Systems. https://www.rts.com/resources/guides/food-waste-america/

Sailing with Bloom. (2017, July 29). *Laundry without a machine.* Sailing with Bloom. https://sailingwithbloom.com/laundry-without-a-machine/

Saltwater FAQs. (2024). NOLA Ready. https://ready.nola.gov/incident/saltwater-intrusion/saltwater-faqs/

Sappo Hill Soapworks. (n.d.). *Activated charcoal history and benefits.* Sappo Hill. https://www.sappohill.com/pilot.asp?pg=Activated_Charcoal

Schleeter, R. (2022, October 24). *The great green wall.* National Geographic Society. https://education.nationalgeographic.org/resource/great-green-wall/

Searle, L., Swainston, D., Murton, B., & Crossley, H. (2023, February 22). *"Should I use pasta water for my plants?": Our gardening experts dish the sauce.* Homes and Gardens. https://www.homesandgardens.com/gardens/should-i-use-pasta-water-for-my-plants

Shallcross, L. (2023, October 17). *How to make a homemade water filter (4 easy DIY projects).* Water Filter Guru. https://waterfilterguru.com/how-to-make-a-homemade-water-filter/

Shrestha, L., Shrestha, R., & Spuler, D. (n.d.). *Boiling - Find tools for sustainable sanitation and water management!* Sustainable Sanitation Water Management. https://sswm.info/sswm-university-course/module-6-disaster-situations-planning-and-preparedness/further-resources-0/boiling

Slater, D. (2019, October 23). *Water scarcity in Brazil.* United States Marine Corps University. https://www.usmcu.edu/Outreach/Marine-Corps-University-Press/Expeditions-with-MCUP-digital-journal/Water-Scarcity-in-Brazil/

Sloan, E. (2021, April 29). *Water test kits promise to tell you if your tap water is safe—but do they work?* Prevention. https://www.prevention.com/health/a22130040/water-test-kits/

Sommerlad, J. (2018, August 29). *How to survive being stranded in the desert.* The Independent. https://www.independent.co.uk/news/world/water-desert-how-survive-cacti-roots-dew-rain-mountains-a8512256.html

Spain, J. (2016, February 25). *25 ways to save water.* Volusia County Florida. https://www.volusia.org/services/growth-and-resource-management/environmental-management/sustainability-and-resilience/water-conservation/25-ways-to-save-water.stml

Staubs, C., & Michelet, J. (2014, July). *Oral rehydration solutions ORS made at home.* Rehydrate. https://rehydrate.org/solutions/homemade.htmStricklin, T. (2022, July 27). *How do cities treat drinking water? (Municipal water treatment explained).* SpringWell Water Filtration Systems. https://www.springwellwater.com/how-do-cities-treat-drinking-water-explained/Stylnpzzalvr. (2007, June 4). *The best rain barrel for less than $15, and where to find a barrel.* Instructables. https://www.instructables.com/How-to-make-a-rain-barrel-1/Tanks for Everything. (2022, December 19). *How to keep water in a water tank from freezing.* Tanks for Everything. https://www.tanksforeverything.co.uk/blog/how-to-keep-water-in-a-water-tank-from-freezing

Tappwater. (2022, September 26). *Guide to constructing a water well for living off-grid.* Tappwater. https://tappwater.co/blogs/blog/well-water-construction-filtration-repairs

Taraba, J., & Heaton, L. (1990, September). *IP-6: Water quality in Kentucky: Using activated carbon filters to treat home drinking water.* UK College of Agriculture, Food, and Environment. https://www2.ca.uky.edu/agcomm/pubs/ip/ip6/ip6.htm

Theodore Roosevelt quote. (n.d.). A-Z Quotes. https://www.azquotes.com/quote/1057462

The Survival University. (2024, February 20). *CDC backcountry water.* The Survival University. https://thesurvivaluniversity.com/survival-tips/wilderness-survival-tips/all-things-water/cdc-backcountry-water/

TMC Health. (2019, May 28). *Is my child dehydrated?* Tuscon Medical Center. https://www.tmcaz.com/blogs/medical-services/childrens-health/is-my-kid-dehydrated

UCLA Health. (2022, June 17). *15 foods that help you stay hydrated.* UCLA Health. https://www.uclahealth.org/news/article/15-food-that-help-you-stay-hydrated

UNCCD. (2023, March 22). *Finding hope in the face of drought in Bol, Chad.* United Nations Convention to Combat Desertification. https://www.unccd.int/news-stories/special-feature/finding-hope-face-drought-bol-chad

UNESCO. (2016, March 29). *The global water quality challenge & SDGs.* UNESCO. https://en.unesco.org/waterquality-iiwq/wq-challenge

UVA Nutrition. (2023). *Homemade oral rehydration solutions.* https://med.virginia.edu/ginutrition/wp-content/uploads/sites/199/2023/12/Homemade-Oral-Rehydration-Solutions-11-2023.pdf

Valley Food Storage. (2023, June 7). *How to desalinate water | A guide on how to desalinate seawater for survival.* Valley Food Storage. https://valleyfoodstorage.com/blogs/inside-vfs/how-to-desalinate-water-survival-desalination

Velayutham, T. (2017). *The importance of water in our daily lives.* India Home Healthcare. https://www.indiahomehealthcare.com/blogpost/the-importance-of-water-in-our-daily-lives/

Walter, J. (n.d.). *How to keep emergency water stored outdoors from freezing.* Super Prepper. https://www.superprepper.com/how-to-keep-emergency-water-stored-outdoors-from-freezing/

Wasson, S. (2022, August 24). *States where it is illegal to collect rainwater.* Today's Homeowner. https://todayshomeowner.com/gutters/guides/states-where-it-is-illegal-to-collect-rainwater/

Water Science School. (2019, November 13). *How much water is there on Earth?* U.S. Geological Survey. https://www.usgs.gov/special-topics/water-science-school/science/how-much-water-there-earth

Watts, J. (2017, November 30). *Brazil's worst drought in history prompts protests and blackouts.* The Guardian. https://www.theguardian.com/world/2015/jan/23/brazil-worst-drought-history

WebMD. (2017, May 26). *What is dehydration? What causes it?* WebMD. https://www.webmd.com/a-to-z-guides/dehydration-adults

Wergin, A. (2022, September 29). *Water is essential to your body.* Mayo Clinic. https://www.mayoclinichealthsystem.org/hometown-health/speaking-of-health/water-essential-to-your-body

Westerlund, S. D. (2023, July 21). *Types of rain barrels [2024 Guide].* Angi. https://www.angi.com/articles/types-of-rain-barrels.htm

WFA. (2017, January 9). *A brief history of the evolution of water filters.* Water Filters Australia. https://www.wfa.com.au/water-filter-evolution/

Whelan, C. (2019, May 13). *Electrolytes food: 25 foods for electrolyte function and more.* Healthline. https://www.healthline.com/health/fitness-nutrition/electrolytes-food#what-are-electrolytes

WHO. (2022, March 30). *Cholera.* World Health Organization. https://www.who.int/news-room/fact-sheets/detail/cholera

wikiHow. (2011, March 25). *Make a water filter.* WikiHow. https://www.wikihow.com/Make-a-Water-Filter

Williams, M.-J. (2023, October 4). *Experts say don't pre-rinse your dishes. We put that advice to the test.* Washington Post. https://www.washingtonpost.com/home/2023/10/04/testing-rinsing-dishes-before-dishwasher/

Woodard, J. (2019, April 8). *Activated carbon filters 101.* Fresh Water Systems. https://www.freshwatersystems.com/blogs/blog/activated-carbon-filters-101

Woodard, J. (2024, January 23). *Reverse osmosis vs distilled water.* Fresh Water Systems. https://www.freshwatersystems.com/blogs/blog/reverse-osmosis-vs-distilled-water

Woods, B. (2024, March 22). *How much does a reverse osmosis system cost? (2024 guide).* This Old House. https://www.thisoldhouse.com/plumbing/reviews/reverse-osmosis-system-cost

WWF. (n.d.). *Recycling greywater for gardening.* World Wildlife Federation. https://wwf.panda.org/discover/knowledge_hub/teacher_resources/project_ideas/recycling/

Zawalsky, B. (2016, March 6). *Water balance: A key to survival.* Boreal. https://boreal.net/articles/survival-training/water-balance.php

ZeroB. (2024, January 30). *Exploring the history of water purification.* Zerob Online.

https://www.zerobonline.com/exploring-the-history-of-water-purification-from-ancient-techniques-to-modern-innovations/

Also by Hadassah Lewis

Discover the secret garden of herbal remedies and how natural medicine can minimize stress and transform your health forever!

Do you cringe every time you take a pharmaceutical drug, worried about the ingredients and side effects?

Have you wondered if there are natural ways to overcome fatigue, relieve pain, boost metabolism, and treat minor ailments and live life to the full?

Discover the art of foraging as you develop the skills and confidence to distinguish edible wild plants from the dangerous ones.

- **Are you weary of questionable store-bought produce?**
- **Would you like to embrace a more self-reliant lifestyle?**
- **Looking for ways to improve your physical, mental, and emotional well-being?**

Foraging offers an opportunity to slow down, enjoy the beauty of nature, and spend time with friends and family as you find wild plants for your next meal or snack.

Get your copy of **31 Edible Wild Plants Absolutely Anyone Can Find in North America: A Beginner's Guide to Safe and Successful Foraging**

This beginner-friendly foraging guide provides the information you need to safely identify edible plants out in the wild. You will discover:

- 31 common edible wild plants with detailed illustrations so you can identify them.
- Insights into the nutritional value of foraged plants.
- Practical tips for safe and successful foraging, with guidance on proper harvesting.

It's time to trade grocery store aisles for open fields and forests as you learn about all the nutritious and delicious plants you can find for yourself.

Scan the QR code or Click below to find the book on Amazon, and get ready to set out on the wonderful path of foraging adventures!

31 Edible Wild Plants Absolutely Anyone Can Find in North America: A Beginner's Path to Safe and Successful Foraging

Made in the USA
Monee, IL
14 September 2026

62974530R10075